Studies in Systems, Decision and Control

Volume 11

Series editor

Janusz Kacprzyk, Polish Academy of Sciences, Warsaw, Poland
e-mail: kacprzyk@ibspan.waw.pl

About this Series

The series "Studies in Systems, Decision and Control" (SSDC) covers both new developments and advances, as well as the state of the art, in the various areas of broadly perceived systems, decision making and control- quickly, up to date and with a high quality. The intent is to cover the theory, applications, and perspectives on the state of the art and future developments relevant to systems, decision making, control, complex processes and related areas, as embedded in the fields of engineering, computer science, physics, economics, social and life sciences, as well as the paradigms and methodologies behind them. The series contains monographs, textbooks, lecture notes and edited volumes in systems, decision making and control spanning the areas of Cyber-Physical Systems, Autonomous Systems, Sensor Networks, Control Systems, Energy Systems, Automotive Systems, Biological Systems, Vehicular Networking and Connected Vehicles, Aerospace Systems, Automation, Manufacturing, Smart Grids, Nonlinear Systems, Power Systems, Robotics, Social Systems, Economic Systems and other. Of particular value to both the contributors and the readership are the short publication timeframe and the world-wide distribution and exposure which enable both a wide and rapid dissemination of research output.

More information about this series at http://www.springer.com/series/13304

Jens Spehr

On Hierarchical Models for Visual Recognition and Learning of Objects, Scenes, and Activities

 Springer

Jens Spehr
Institut für Robotik und Prozessinformatik
Technische Universität Braunschweig
Braunschweig
Germany

ISSN 2198-4182   ISSN 2198-4190   (electronic)
Studies in Systems, Decision and Control
ISBN 978-3-319-11324-1   ISBN 978-3-319-11325-8   (eBook)
DOI 10.1007/978-3-319-11325-8

Library of Congress Control Number: 2014949167

Springer Cham Heidelberg New York Dordrecht London

Printed on acid-free paper

[Springer International Publishing AG Switzerland] is part of Springer Science+Business Media
(www.springer.com)

Preface

The understanding of images or image sequences is a very challenging task. The main advantage that image data is rich in information is simultaneously the main problem since important has to be separated from unimportant information. This separation addresses more than just the classical foreground background segmentation, where the color image is generally transformed into a binary image. It also refers to pose estimation, where the color image is transformed into a few pose parameters, or to anomaly detection, where the color image is transformed into binary information. This book shows how these challenging tasks can be solved even for real-time applications.

The main part of this book was developed during my activity as a research assistant at the "Institut für Robotik und Prozessinformatik" at the "Technische Universität Carolo-Wilhelmina zu Braunschweig". First and foremost, I am deeply grateful to Prof. Dr. Friedrich M. Wahl. He gave me the opportunity to develop own ideas, to prove them within projects, to discuss them, and finally to write them down. I also want to acknowledge the valuable help of colleagues, reviewers, and students. In particular, I would like to thank Prof. Dr. Joachim M. Buhmann, Simon Winkelbach, Markus Rilk, Dennis Rosebrock, Volker Schomerus, Carsten Last, Paulin Pekezou Fouopi, and Stephan Kluge.

Parts of this work are closely connected with projects. I thank therefore the Group Research at Volkswagen AG for the project "Camera-based measurement of parallel and perpendicular parking spaces" as well as the Lower Saxony Ministry of Science and Culture for support of the research network "Design of Environments for Ageing" through the "Niedersächsisches Vorab" grant program (grant ZN 2701). For the successful and efficient teamwork within these projects I would like to thank Daniel Mossau, Richard Auer, Matthias Gietzelt and Mehmet Gövercin.

Finally, I wish to thank my wife for her continued love and support. Without her untiring patience I could never have done this.

Salzgitter   *Jens Spehr*
September, 2014

Contents

List of Figures

List of Tables

Chapter 1
Introduction

The visual recognition and learning of objects, scenes, and activities is a highly relevant but simultaneously very difficult task, which has applications in many areas such as robotics, industry, consumer electronics, aerospace, transportation systems, or ambient assistant living. Just a decade ago, computer vision applications were mainly limited to machine vision, where the actual learning and recognition task were strongly restricted due to simplified environment conditions. But the more powerful new hardware becomes, the more are new approaches able to integrate uncertainties as well as dependencies between objects and a scene, allowing them to act in real world scenarios. Even approaches already proposed in the past, which were up to now unthinkable to calculate in real-time, are more and more revived.

In industrial scenarios, the standard computer vision task of detecting and recognizing objects can often typically be solved using a simple threshold method and appropriate shape descriptors. In real-world scenes, however, multiple views, different articulations, and scales have to be regarded by combining different appearance cues in order to get a robust detector. Furthermore, the representation should provide an invariance against the typical image processing issues such as illumination changes, background clutter, or occlusions. The detected objects can for example be used for content-based image retrieval or in an augmented reality application where the view of a real-world environment is augmented by computer-generated graphics. Other currently very interesting examples are ambient assisted living applications. They are interesting particularly with regard to our aging society, which makes the support of elderly people in their home environments more and more necessary. On the one hand the number of elderly people increases, and on the other hand the number of people being able to care about them decreases. As a result of this decreasing number of nursing staff, free places in residentials or nursing homes become scarce. One promising way to solve this dilemma is to allow elderly people to stay as long as possible in their familiar home environment while ensuring the health of the person by means of a monitoring system. This monitoring system extracts estimates of vital signs such as gait speed and detects critical values as well as critical changes. Vision-based sensors like cameras have the advantage of being unintrusive, invisible for the user, and need no interaction between the user and the system. The system

© Springer International Publishing Switzerland 2015
J. Spehr, *On Hierarchical Models for Visual Recognition & Learning of Objects, Scenes, & Activities,*
Studies in Systems, Decision and Control 11, DOI: 10.1007/978-3-319-11325-8_1

can be used to detect suddenly happening critical events like e.g. falls. A vision sensor allows this detection but needs an efficient image processing framework that extracts features like body hight or body axis from the image data [220]. Vision-based approaches can also be used for fall prevention. By means of a gait analysis, features like step width, step height, speed and speed variance, gait harmony, compensation motion and body sway can be extracted and used to describe how critical the current gait parameters are [219]. Even more challenging is the representation of human behavior patterns and the detection of anomalies, which might be indicating emerging health problems.

To give a further promising application area, autonomous driving of an intelligent vehicle requires a robust sensory based perception of the environment. Provided by common 3d reconstruction techniques such as structure from motion, a camera poses a very powerful 3d sensor. However, the interpretation of the gathered data is still very challenging and demands sophisticated approaches, which provide a high-level understanding of the scene in terms of the detection and recognition of the road, other cars, pedestrians, parking places and so on.

Although this application examples seem to have nothing in common, their image processing schemes are deeply connected since each application is based on low-level sensor input (gray/color values of each pixel) and tries to derive some high-level interpretation such as the information about the human body, behavior patterns or positions of free parking places. What makes scene interpretation and understanding so difficult are the different abstraction levels underlying an image. The actual vision sensor supports just the lowest level of these abstractions. All higher levels have to be derived based on this information. During the bottom-up processing many assumptions and conditions have to be regarded. Under these assumptions are simplifications which are mainly made in order to reduce the search space and thus to reduce the computational effort. Among theses assumptions are furthermore restrictions of the visual appearance of objects, which occur e.g. due to specific views of an object or specific articulations. Most of the existing approaches try to directly infer high-level information based on the low-level pixel information. Often the subspace and variations of the visual appearance within the borders of the specified variations are modeled with appropriate representations. Representations like the SIFT [135] or HOG [36] feature descriptor try to model these variations by means of histograms which are capturing the statistical variations in spatially arranged cells. Although these representations achieve impressive performances in common benchmark datasets, they are not modeling intermediate hierarchy levels. Therefore, these models are unfeasible for handling multi-view and multi-object representations. Modeling of the intermediate levels can cope with these challenges by sharing object parts or visual primitives. Here, a hierarchical model is particularly suitable for describing the decomposition into parts or primitives. In order to illustrate the hierarchical representation and the idea of sharing let us return to our application examples. Objects can naturally be decomposed into parts, these parts into visual primitives and finally the primitives into local gradients and color values. Instead of modeling the elements (different views or articulations) independently, sharing allows to reduce the total number of parts and primitives. It thus leads to better generalization properties and to an efficient representation, which reduces the computational costs [234]. It also allows hierarchical models to be applied in real-world applications. Since parts and primitives are shared we have to infer information about these elements just once and can make them available for all associated parent nodes. This minimizes redundant calculations and increases the performance of the algorithm. For articulated structures, such as the human body, the degree of sharing is even higher. For instance, if just the left upper arm moves and the other body parts stay constant, we can share the geometrical structure and appearance of the whole body between different configurations and have just to model additionally the different configurations of the left arm. Similar decompositions and sharing properties can be found for activities, which can be decomposed in actions, actions again in action primitives, and also for traffic scenes, which can be decomposed in objects like cars and pedestrians, and this objects again in simple primitives.

1.1 Hierarchical Models: Objectives and Challenges

In many applications, an efficient representation of objects, scenes, and activities is needed. In order to meet real-time requirements and to work robustly many applications introduce boundary conditions and thus simplify the problem. Generally, these approaches are by far not reaching the performance of more recently published methods and models like hierarchical representations which achieve brilliant results on common challenging benchmark datasets. The goal of this work is to develop an efficient hierarchical representation of objects, scenes, and activities, which is attractive for many practical applications. Building on the excellent results of hierarchical approaches, we develop efficient learning and inference methods, which increase the overall performance and make our approach real-time applicable.

Furthermore, we aim at applying our hierarchical representation to datasets, which just contain few object instances and thus differ strongly from common datasets. Common benchmark datasets support hundreds of instances for one specific object class and are generally lavishly created or collected. For example in [269], a full time annotation team parsed the structures of 500,000 images covering objects of 280 categories. However, in many applications the detection, recognition and pose estimation of just one specific object is desired. Often only a single instance of one specific view, scale and articulation is given. Or in the borderline case no instances are available, so that manually generated prior knowledge of e.g. spatial arrangements has to be used. Here, especially for problems like pose estimation it is unthinkable to build up datasets containing different instances for every specific view, scale and articulation. Another aspect to consider when learning from few examples is the generalization ability of the underlying models. Based on the instances seen during learning, the model should be able to generalize the representation to new unseen instances during recognition. Here, hierarchical models are well suited as well since they exploit the reusability of parts and primitives leading to better generalization properties.

1.2   Overview of Contributions

The main contributions of our new proposed hierarchical framework are (to the best of our knowledge):

- We introduce a hierarchical representation of view-point dependent instances and directly encode the scale and rotation into the hierarchy. Our hierarchy is less constrained than other representations found in the literature, i.e our hierarchies have up to 20 levels, and flexible inter layer dependencies. This flexibility maximizes the reusability of parts leading to a compact and efficient representation. Differing from previous hierarchical models, where the highest level is explicitly denoted as an object layer, our representation couples the hierarchical level directly with the size and complexity of an object. Thus, a complex object with texture will be represented on a higher level, while a simple one like a circle will be represented on a lower level.
- We combine the hierarchical representation with a coarse-to-fine search by means of a similarity hierarchy. Although the idea of coarse-to-fine searches is well-known, we are the first who directly integrated the idea into a hierarchy at different abstraction levels and combine it with a scale space representation. As we will see, this combination efficiently generates and verifies part and object hypotheses. They are generated in a compositional hierarchy at a coarse scale, refined in a coarse-to-fine similarity hierarchy, and finally evaluated in a compositional hierarchy at a fine scale.
- We propose an unsupervised top-down learning method that maximizes the reusability of parts and supports to learn efficient hierarchies offline as well as online.
- We apply our approach to human pose estimation, activity representation, and scene understanding for intelligent vehicles. The contributions in these fields will be separately highlighted at the beginning of each application chapter.
- Minor contributions are related to our applied inference techniques. We propose to combine the bottom-up message passing with a top-down passing step, perform the bottom-up passing in several sequential sweeps and introduce importance sampling for high-level observable nodes.

1.3   Outline of the Monograph

The remaining chapters are organized as follows:

- **Chapter 2** briefly summarizes the basic concepts of probabilistic graphical models as found in the open literature. The chapter covers a general overview and highlights particular elements and mechanisms of the hierarchical models. First, a survey of the basic concepts in graph theory is presented in which important terms and definitions from graph theory are emphasized. After that, we will take a closer look at undirected and directed graphs and explain the basic concept of Markov random fields. Finally, probabilistic basics and inference methods (belief propagation and nonparametric belief propagation) are explained and illustrated by means of a simple hierarchical model.

- **Chapter 3** presents our new hierarchical graphical models. We start with a formal definition of our hierarchies. The related work is divided into different models (part-based models, constellation models, compositional hierarchies) and also into related principles (feature and part sharing, coarse-to-fine hierarchies, biologically inspired models). After a simple example of a compositional hierarchy we clarify which information sources a node in the hierarchy can access, how observations can be fed into the hierarchy and how spatial and temporal dependencies are included in the representation. In the section about our new compositional hierarchical model, we describe how different scales, orientations and views can be efficiently represented, how similarities can be used to share information and how the compositional hierarchy can be combined with a coarse-to-fine hierarchy at different abstraction levels and scales. Finally, new efficient inference approaches will be proposed.

- **Chapter 4** is dedicated to the learning of hierarchical models. It starts with a comparison of different common benchmark datasets and the datasets used in this monograph. After the presentation of related work, which is divided into bottom-up and top-down approaches, we introduce our new top-down learning approaches (offline and online) including structure and parameter learning. Finally, we describe how the scale and rotation invariant representation is achieved.

- **Chapter 5** is dedicated to object recognition. Since related hierarchical models were already summarized in Chap. 3, we focus on the low-level feature extraction techniques known from literature in the related work section. We then explain in more detail the low-level features used in this monograph, and study the different learning approaches (offline and online) and inference techniques (combined bottom- up and top-down propagation, number of sweep, coarse-to-fine).

- **Chapter 6** provides application results for human pose estimation. We briefly summarize related works and explain the idea of hierarchical decomposition of the human body. After describing the learning and design of the hierarchies, we demonstrate their capability by means of a gait analysis application.

- **Chapter 7** describes the hierarchical representation of human behavior patterns. We give an overview of related low-level action primitives as well as related high-level action models. After describing the idea of our new hierarchical action representation and of our optical flow primitives, we describe the unsupervised learning and highlight differences to previously described approaches. We then demonstrate the application of the proposed hierarchical graphical models using a dataset containing hand movements, where we show how sequences of actions can be modeled. Furthermore, we use a dataset containing activities of daily living where supervised learning of activities is intractable due to the large amount of video data. We show how the proposed unsupervised learning approach can automatically extract motion patterns at different hierarchy levels corresponding to different complexities.

- **Chapter 8** applies the hierarchical models to scene understanding for intelligent vehicles. After summarizing related work, which we divide into general scene understanding approaches and different categories of 3d computer vision approaches, we explain the idea of hierarchical scene interpretation and motivate the application in intelligent vehicles, where erroneous and noisy sensors like cameras are used. We describe our new concept of virtual sensors, and introduce some key aspects of the hierarchical representation (sharing parts, reduced particle sets, observable high-level nodes, periodic variables). After a performance and accuracy analysis we demonstrate the hierarchical representation with a parking spot finding application.

- **Chapter 9** concludes this work, summarizes the key ideas of our approach as well as the applicational results, and discusses some extensions and directions for future research.

Chapter 2
Probabilistic Graphical Models

In this chapter, we will briefly summarize the basic concepts of probability as well as graph theory. We start with important terms and definitions from graph theory and emphasize the relation to our hierarchical models. Directed and undirected graphs will be introduced and compared with each other. Then, we will give an overview of the random variables and the underlying probability distributions concentrating on nonparametric distributions and methods such as kernel density estimation. Finally, a simple hierarchical model is used to explain and illustrate inference methods.

2.1 Brief Review of Probabilistic Graphical Models

Probabilistic graphical models are well suited to describe the computer vision, machine learning and pattern recognition tasks investigated in this monograph. They allow to decompose complex multivariate joint distributions over various variables into a product of smaller and simpler subsets of these variables. This decomposition is done by a graph capturing the conditional independencies among the variables. The standard example to illustrate this is the multimodal model of the joint distribution $p(x_1, ..., x_N)$ over binary variables x_i [226, 16]. Without using independence properties the model would require a table with 2^N entries and the computation of the marginal $p(x_1)$ would need summing over the 2^{N-1} binary states. Even for a small number of variables ($N \approx 100$) the storage and manipulation of this density is clearly infeasible. The introduction of conditional independencies constrains the interaction between the variables, which leads to a factorization of the joint distribution and thus to efficient algorithms. E.g. the joint distribution of a chain graph can be factorized as $p(x_1, ..., x_N) \propto \prod_{i=1}^{N-1} \psi(x_i, x_{i+1})$ (see Fig. 2.1 for a comparison of a fully connected and a chain graph) . Before explaining the factorization properties in detail we will briefly review some definitions from graph theory.

© Springer International Publishing Switzerland 2015
J. Spehr, *On Hierarchical Models for Visual Recognition & Learning of Objects, Scenes, & Activities,*
Studies in Systems, Decision and Control 11, DOI: 10.1007/978-3-319-11325-8_2

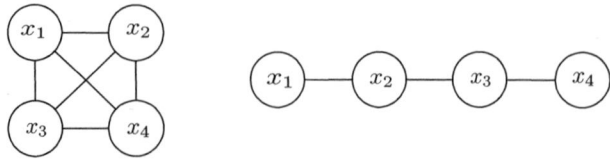

Fig. 2.1 Examples of probabilistic graphical models with four nodes: (a) Fully connected graph with $p(x_1, x_2, x_3, x_4) \propto \psi(x_1, x_2, x_3, x_4)$. (b) Chain graph $p(x_1, x_2, x_3, x_4) \propto \psi(x_1, x_2)\psi(x_2, x_3)\psi(x_3, x_4)$.

2.1.1 Survey of the Basic Concepts in Graph Theory

The graph $\mathcal{G} = (\mathcal{V}, \mathcal{E})$ comprises a set $\mathcal{V}$ of *nodes* (also called vertices) together with a set $\mathcal{E}$ of *edges* (also called links or arcs) [119, 82, 112, 10, 226]. Two nodes $i, j \in \mathcal{V}$ are connected by an edge $(i, j) \in \mathcal{E}$, where the edges may be directed (asymmetric) or undirected (symmetric). In the directed case, an edge (i, j) has a particular directionality depicted by an arrow pointing from the *parent* node i to the *child* node j (see Fig. 2.2(a)). For node j the set of all parents $\Gamma(j)$ is given by $\Gamma(j) = \{i \in \mathcal{V} | (i, j) \in \mathcal{E}\}$ and the set of all children $\Xi(j)$ is given by $\Xi(j) = \{k \in \mathcal{V} | (j, k) \in \mathcal{E}\}$. The set of directed *neighbors* of node j can then be defined as $\Upsilon(j) = \Xi(j) \bigcup \Gamma(j)$. In the undirected case, the edge (i, j) has no orientation, so that it is identical to the edge (j, i) and it is depicted by an arrowless line (see Fig. 2.2(b)). The *neighbors* of j, $\Upsilon(j)$, are those nodes directly connected to j. A *clique* is a fully connected subset of nodes for which all members are neighbors. A clique is called a *maximal clique* if it is not contained in any other clique in $\mathcal{G}$. Thus, it is not possible to add any other node from the graph, which is connected to all members of the clique, to a maximal clique. A *path* from node i to node j is a sequence of nodes such that from each of its nodes there is an edge to the next node in the sequence. If a directed path starts and ends with the same node the path is also called a *cycle*. A graph is acyclic if it has no cycles, so that one cannot move from one node to another node according to the edge direction and end at the starting node. Similarly, a *loop* is a path with more than two nodes that starts and returns to the same node regardless of the edge direction. An undirected as well as a directed graph is *singly connected* if there is only one path from any node i to any other node j. A singly connected graph is also called a *tree* and a disjoint union of trees is called a *forest*. A *leaf* of a singly connected graph is a node that is connected to exactly one other node.

2.1.2 Undirected Graphical Models

A graphical model associates each node $i \in \mathcal{V}$ of a graph $\mathcal{G} = (\mathcal{V}, \mathcal{E})$ with a random variable x_i taking values in the sample space $x_i \in \mathcal{X}_i$ and each edge $e_i \in \mathcal{E}$ with a probabilistic interaction between the neighboring variables. Therefore, a graphical model can be seen as a joint probability distribution $p(x)$ encoding the independence structure between a set of random variables $x = \{x_1, ..., x_N\}$ by means of graph $\mathcal{G}$. The local dependencies or constraints between the variables are defined by *potentials*, where the potential $\psi(x)$ is a non-negative function of the variable x, $\psi(x) \geqslant 0$. A non-negative function of a set of variables $\psi(x_1, ..., x_N)$ is also called a joint potential. A potential satisfying $\sum_x \psi(x) = 1$ represents a distribution. Undirected graphical models are also referred to as *Markov random fields* (MRF) or *Markov networks* since they follow the Markov properties, which can be divided into global, local and pairwise Markov properties [10, 226]. Consider an undirected graph $\mathcal{G} = (\mathcal{V}, \mathcal{E})$ and let f, g and h denote three disjoint subsets of $\mathcal{V}$. The set h separates the set f from the set g if every path from f to g passes through h. Thus, if f and g are not connected by an edge, then f is separated from g. If h separates f from g the *global Markov property* states that the set of variables $\boldsymbol{x}_g$ and $\boldsymbol{x}_f$ are independent conditioned on the set of variables $\boldsymbol{x}_h$

$$p(\boldsymbol{x}_f, \boldsymbol{x}_g | \boldsymbol{x}_h) = p(\boldsymbol{x}_f | \boldsymbol{x}_h) p(\boldsymbol{x}_g | \boldsymbol{x}_h) \tag{2.1}$$

The *local Markov property* implies that x_i is conditionally independent of all other variables $\boldsymbol{x}_{\mathcal{V} \setminus i}$ given its neighbors $\boldsymbol{x}_{\Upsilon(i)}$

$$p(x_i | \boldsymbol{x}_{\mathcal{V} \setminus i}) = p(x_i | \boldsymbol{x}_{\Upsilon(i)}) \tag{2.2}$$

A graph has the *pairwise Markov property* for all non-adjacent nodes i and j if

$$p(x_i, x_j | \boldsymbol{x}_{\mathcal{V} \setminus \{i,j\}}) = p(x_i | \boldsymbol{x}_{\mathcal{V} \setminus \{i,j\}}) p(x_j | \boldsymbol{x}_{\mathcal{V} \setminus \{i,j\}}) \tag{2.3}$$

For positive potentials the global, local and pairwise Markov properties are equivalent. The Markov properties can now be used to formulate a decomposition of the joint distribution. The *Hammersley-Clifford theorem* states that, given an undirected graph $\mathcal{G}$ and a set of cliques $\mathcal{C}$, any strictly positive density that satisfies one of the Markov properties with respect to $\mathcal{G}$ can be factorized over the cliques $\mathcal{C}$

$$p(\boldsymbol{x}) = \frac{1}{Z} \prod_{c \in \mathcal{C}} \psi_c(x_c) \tag{2.4}$$

where

$$Z = \sum_x \prod_{c \in \mathcal{C}} \psi_c(x_c) \tag{2.5}$$

is a normalization constant known as the partition function. The functions ψ_c are often referred to as *clique potentials*. The Hammersley-Clifford theorem is fundamental since it describes how a MRF is parameterized by potential functions on the cliques of the corresponding undirected graph. Of a special kind are those MRFs where the potential functions are just defined between neighboring nodes

$$p(\boldsymbol{x}) \propto \prod_{(i,j) \in \mathcal{E}} \psi_{ij}(x_i, x_j) \tag{2.6}$$

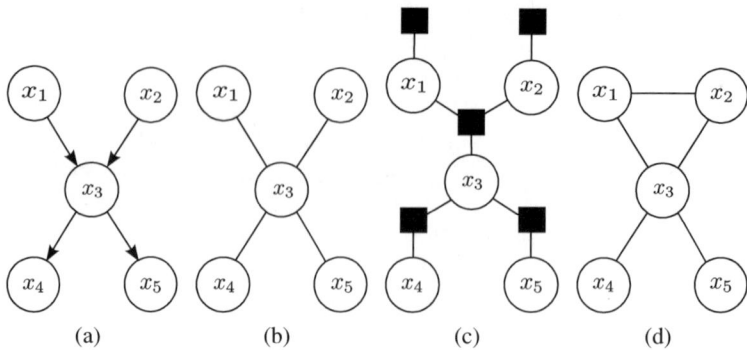

Fig. 2.2 Graphical probability models. Please note that all graphs except (b) represents the same distribution over the random variables $\{x_1, ..., x_5\}$ (see [115, 226]). (a) A Bayesian network. (b) A Markov random field. (c) A factor graph. (d) A moral graph.

In this case the Hammersley-Clifford theorem guarantees that the MRF is Markov with respect to $\mathcal{G}$ since each pair of neighboring nodes defines a clique. Due to the pairwise defined potentials the corresponding MRFs are also called *pairwise Markov random fields*. An example is given in Fig. 2.2(b). Here, the pairwise clique potentials allows us to factorize the joint distribution as

$$p(\boldsymbol{x}) = \frac{1}{Z}\psi_{13}(x_1, x_3)\psi_{23}(x_2, x_3)\psi_{34}(x_3, x_4)\psi_{35}(x_3, x_5) \tag{2.7}$$

where Z is again the normalization constant. Since the potentials in eq. 2.6 are just capturing the pairwise dependencies between neighboring nodes, it is often convenient to extended it by using additional single-node potentials $\psi_i(x_i)$ for each node $i \in \mathcal{V}$ [226]

$$p(\boldsymbol{x}) \propto \prod_{(i,j)\in\mathcal{E}} \psi_{ij}(x_i, x_j) \prod_{i\in\mathcal{V}} \psi_i(x_i) \tag{2.8}$$

Factor graphs are another type of undirected graphs that are mainly used to design efficient inference algorithms [115]. Formally a factor graph is a *hypergraph* where *hyperedges* are connecting subsets with more than two nodes. In contrast to MRFs the potentials are not defined on the cliques of the graph. Instead, a factor graph uses the hyperegdes to represent the potentials $\psi_f(x_f)$ as a function of the corresponding variables x_f. Thus, a factor graph defines the joint distribution as

$$p(\boldsymbol{x}) \propto \prod_{f\in\mathcal{F}} \psi_f(x_f) \tag{2.9}$$

where the variable x_f and the potentials ψ_f can be represented as nodes of a *bipartite graph*. An example is shown in Fig. 2.2(c) with a circular node for each variable and a square node for each hyperedge.

2.1.3 Directed Graphical Models and Their Relation to Undirected Graphical Models

Directed graphical models are based on graphs $\mathcal{G} = (\mathcal{V}, \mathcal{E})$ without directed cycles, so-called *directed acyclic graphs* [16, 10, 226]. The corresponding models are also referred to as *Bayesian networks* or *Belief networks*. Similar to undirected graphical models, each node $i \in \mathcal{V}$ in a Bayesian network represents a random variable x_i. While in an undirected graphical model potential functions define the local interaction between the associated nodes, a directed model captures the causal dependencies by means of conditional density distributions. The joint distribution of a Bayesian network factorizes as

$$p(\boldsymbol{x}) = \prod_{i \in \mathcal{V}} p(x_i | \boldsymbol{x}_{\Gamma(i)}) \qquad (2.10)$$

where $p(x_i | \boldsymbol{x}_{\Gamma(i)})$ is the conditional density of node x_i conditioned on its parent nodes $\boldsymbol{x}_{\Gamma(i)}$. In order to illustrate the relationship to an undirected graph, let us consider the directed graph in Fig 2.2(a). The joint distribution is given by

$$p(\boldsymbol{x}) = p(x_1)p(x_2)p(x_3|x_1, x_2)p(x_4|x_3)p(x_5|x_3) \qquad (2.11)$$

where the nodes that have no parent ($\Gamma(i) = \emptyset$) are written as $p(x_i|\boldsymbol{x}_{\Gamma(i)}) = p(x_i)$, here $p(x_1)$ and $p(x_2)$ [226]. Let us now consider an undirected graph with the same graph structure, where each directed edge is simply replaced by an undirected one (see Fig 2.2(b)). Comparing the joint density of the direct model to those of the undirected, see eq. 2.7, gives us $\psi_{34}(x_3, x_4) = p(x_4|x_3)$ and $\psi_{35}(x_3, x_5) = p(x_5|x_3)$. However, one problem arises concerning the product $\psi_{13}(x_1, x_3)\psi_{23}(x_2, x_3)$ that can, in general, not equal $p(x_1)p(x_2)p(x_3|x_1, x_2)$. To solve this problem the set of variables $\{x_1, x_2, x_3\}$ must all belong to a single clique. This can only be ensured by adding an extra undirected edge between the parents of the node x_3 (see Fig. 2.2(d)). The step of adding undirected edges between the parents of a node is also known as *moralization* since the parents are "married" and the corresponding graphs are called *moral graphs*. In this case, $p(\boldsymbol{x})$ is given by

$$p(\boldsymbol{x}) = \frac{1}{Z}\psi_{123}(x_1, x_2, x_3)\psi_{34}(x_3, x_4)\psi_{35}(x_3, x_5) \qquad (2.12)$$

As can be seen, now $\psi_{123}(x_1, x_2, x_3) = p(x_1)p(x_2)p(x_3|x_1, x_2)$ and $Z = 1$. It is worth noting that after moralization the variables $\{x_1, x_2, x_3\}$ are fully connected and that thus the conditional independence properties of the directed graph are lost. While in an undirected graph the local Markov property (see eq. 2.2) guarantees that a node is conditionally independent of the remaining nodes given its neighbors, in a directed graph a node is isolated from the rest of the graph by its *Markov blanket* [179]. The Markov blanket or *Markov boundary* of a node x_i comprises the parents $\boldsymbol{x}_{\Gamma(i)}$, the children $x_{\Xi(i)}$ and the children's parents $x_{\Gamma(j)}$ for $j \in \Xi(i)$. The children's parents $x_{\Gamma(j)}$ are necessary since they can be used to explain away [1] node x_i.

2.1.4 Observable Nodes

In the previous section, we showed how local interactions between random variables can be used to express complex joint probability distributions of an arbitrary number of variables. So far, we have considered all these random variables x to be *hidden* oder *latent*. For inference tasks this encoding of structural dependencies is particularly attractive if for some of the nodes in the graph observed values are available. We will consider observed values to be noisy observations y that are connected to hidden random variables x. These observations will be depicted as shaded nodes. Commonly, we are interested in the *posterior* distribution $p(x|y)$ of the latent variables x given their observed values y. Using Bayes' theorem we can calculate the posterior as

$$p(x|y) = \frac{p(y|x)p(x)}{p(y)} \tag{2.13}$$

where the *likelihood* $p(y|x)$ gives the probability that the variables x generates the observed values y. The likelihood does not have to represent a probability distribution over x, so that the integral $\int p(y|x)dx$ does not have to equal one. The marginal distribution $p(x)$ holds information about x before the measurement is captured and it is called the *prior* over the latent variables x. The *evidence* or *marginal likelihood* $p(y)$ acts like a normalization term and is often omitted since it is independent of x and can be calculated as

$$p(y) = \int_{\mathcal{X}} p(y|x)p(x)dx \tag{2.14}$$

In a pairwise MRF it is often convenient to combine the observations with the single-node potential $\psi_i(x_i)$. We can then express the posterior distribution of a pairwise MRF using eq. 2.8 as

$$p(x|y) = \frac{p(x,y)}{p(y)} \propto \prod_{(i,j)\in\mathcal{E}} \psi_{ij}(x_i, x_j) \prod_{i\in\mathcal{V}} \psi_i(x_i, y) \tag{2.15}$$

If the observations can be decomposed into local measurements $y_i \in y$, which are directly connected to associated latent nodes x_i, we can also write the single-node potential as $\psi_i(x_i, y) = \psi_i(x_i, y_i)$ [226].

---

[1] The term "explaining away" is often used in the literature to describe the influence of the children's parent on a node.

2.2 Probability Distributions

We will now have a closer look at the random variables x_i and the probability distributions $p(x_i)$ underlying them. Furthermore, we will assume that N independent samples $x_i^{(n)} \sim p(x_i)$ drawn from the distribution are given. An example of a 2d probability distribution can be seen in Fig. 2.3(a) with the corresponding set of samples in Fig. 2.3(b). There are two main classes for modeling probability distributions. *Parametric* distributions are defined by a relative small set of parameters specifying the model function. A typical example is the Gaussian distribution, also known as normal distribution, which is defined by a mean μ and a variance parameter σ^2

$$\mathcal{N}\left(x; \mu, \sigma^2\right) = \frac{1}{\sqrt{2\pi\sigma^2}} \exp\left(-\tfrac{1}{2}\left(\tfrac{x-\mu}{\sigma}\right)^2\right) \qquad (2.16)$$

In many cases, the Gaussian is an appropriate model as long as the associated data is unimodal. However, in more realistic scenarios the model has to represent complex and multimodal forms of a distribution due to outliers, ambiguities, and other non–Gaussian effects (see for example 2.3(c)). *Nonparametric* distributions are making fewer assumption about the distribution enabling them to adapt to non-Gaussian effects. One example of nonparametric approaches is the *histogram* method [16, 215]. Let us consider the continuous random variable x that takes values in the sample space $\mathcal{X}$. The standard histogram approach divides the sample space into distinct bins of width Δ_i. As usual, for frequentist methods, the probability of each bin is then calculated by counting the number of observations of x falling into each of the disjoint bins

$$p_i = \frac{n_i}{N\Delta_i} \qquad (2.17)$$

where N is the total number of observations. An example is shown in 2.3(e). Histogram density models are appropriate methods as long as the dimension d of the variable x and the number of bins M for each dimension is small since the total number of bins grows exponentially according to M^d. Another problems arises concerning the right choice of the width Δ_i. If the width is too large, the resulting distribution is under certain conditions not able to model the multimodal properties of the underlying data. Especially the edge locations can distort the resulting distribution significantly and cause discontinuities. On the other hand, if the width is too small the distribution can become noisy and spiky. *Kernel density estimation* is closely related to the histogram method but it overcomes its drawbacks. Discontinuities are avoided by using symmetric kernel functions $k\left(\cdot\right)$ to smooth the samples, and the locations of the kernels are individually adapted to the data points $x^{(n)}$ and not fixed as the edges of the bins are. The estimated density is calculated as

$$p(x_i) = \frac{1}{N} \sum_{n=1}^{N} \frac{1}{h^d} k\left(\frac{x_i - x_i^{(n)}}{h}\right) \qquad (2.18)$$

where h is a smoothing parameter that is also called the bandwidth. A common choice is the Gaussian kernel

$$p(x_i) = \sum_{n=1}^{N} w_i^{(n)} \mathcal{N}\left(x_i; x_i^{(n)}, \Lambda_i\right) \tag{2.19}$$

where $w_i^{(n)}$ are weights associated to the normal distribution. The bandwidth Λ_i of the density can be estimated via the "rule of thumb" [215] method, which is a simple but robust covariance estimate. The method assumes that the density is Gaussian and estimates the optimal bandwidth as

$$\Lambda_{\text{opt}} = \left(\frac{4}{d+2}\right)^{\frac{1}{d+4}} \sigma^2 N^{-\frac{1}{d+4}} \tag{2.20}$$

where d is the dimension of the random variable and σ^2 can e.g. be chosen as the average marginal variance $\sigma^2 = d^{-1} \sum_i \sigma_{ii}$ of a general covariance matrix with diagonal elements σ_{ii}. As can be seen in Fig. 2.3(f) this method is able to keep the multimodal properties of the density but tends to over smooth the distribution. In many applications it is convenient to represent the density function directly by the set of discrete samples

$$p(x_i) \approx \sum_{n=1}^{N} w_i^{(n)} \delta(x_i, x_i^{(n)}) \tag{2.21}$$

Here, $w_i^{(n)}$ are again weights associated to the samples and that fulfill $\sum_{n=1}^{N} w_i^{(n)} = 1$. This weighted sample set $\{(x_i^{(n)}, w_i^{(n)})\}_{n=1}^{N}$ is commonly used in the popular particle filter method. The samples are therefore often also referred to as *particles*. A kernel density estimate of a particle set can simply be derived by convolution with a Gaussian kernel with its covariance determined by the rule of thumb.

2.3   Inference in Graphical Models

As mentioned in Sec. 2.1.4 the problem of inference in graphical models is fundamental whenever noisy observations of random variables are available and the belief of the latent variables has to be calculated. In this section, we will briefly summarize some important aspects of inference in graphical models and we will prepare some basic algorithms, namely the belief propagation algorithm, necessary for the inference algorithms in the following chapters. Let us assume, we want to find the posterior marginal distribution $p(x_i|\boldsymbol{y})$ for a specific node i given global observations $\boldsymbol{y}$. The marginal distribution is obtained by

$$p(x_i|\boldsymbol{y}) = \int_{\mathcal{X}_{\mathcal{V}\setminus i}} p(\boldsymbol{x}|\boldsymbol{y}) d\boldsymbol{x}_{\mathcal{V}\setminus i} \tag{2.22}$$

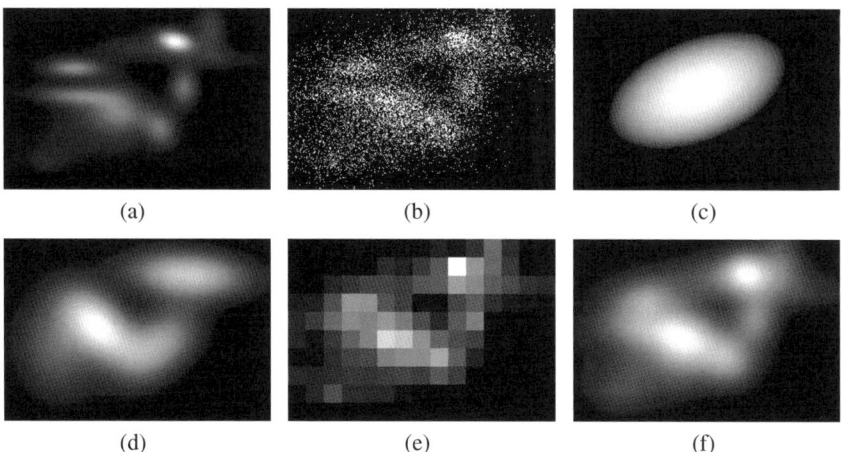

Fig. 2.3 Different representations of a bivariate multimodal density: (a) Ground truth density. (b) Sample set. (c) Gaussian distribution. (d) Mixture of four Gaussians. (e) Histogram density. (f) Kernel density estimate.

where $\mathcal{V}\backslash i$ denotes all nodes of the graph except i. Thus, if we consider a brute-force implementation, the marginal distribution involves a summation over all values of the joint sample space $\mathcal{X}$ except those corresponding to variable x_i. As already mentioned at the beginning of this chapter, even for a small number n of binary variables the joint sample space has 2^N elements and is far too large. If we consider the more general case, where the variables can take one of K discrete values, the computation requires K^N operations [226]. We will now show how efficient methods can be developed by passing belief messages through the graph. Let us consider the tree-structured graph in Fig. 2.4, which will be the primary graph structure for the hierarchies discussed in the following chapters. Suppose that we want to calculate the posterior marginal distribution $p(x_3|\boldsymbol{y})$. Since the tree can be represented by a pairwise MRF we can apply eq. 2.8 to calculate the joint distribution $p(\boldsymbol{x})$. The marginal can then be calculated as

$$p(x_3) \propto \int \cdots \int \psi_1(x_1)\psi_{13}(x_1, x_3)\psi_2(x_2)\psi_{12}(x_1, x_2)\psi_3(x_3)\psi_{24}(x_2, x_4)\psi_4(x_4)$$
$$\cdots \times \psi_{25}(x_2, x_5)\psi_5(x_5)\psi_{36}(x_3, x_6)\psi_6(x_6)\psi_{37}(x_3, x_7)\psi_7(x_7)dx_1 dx_2 dx_4 dx_5 dx_6 dx_7 dx_8$$

After rearranging the order of the integrals and the multiplications the equation can be written as

$$p(x_3) \propto \psi_3(x_3)\left[\int \psi_{36}(x_3, x_6)\psi_6(x_6)dx_6\right]\left[\int \psi_{37}(x_3, x_7)\psi_7(x_7)dx_7\right]$$
$$\cdots \times \left[\int \psi_{13}(x_1, x_3)\psi_1(x_1)\left[\int \psi_{12}(x_1, x_2)\psi_2(x_2)\left[\int \psi_{24}(x_2, x_4)\psi_4(x_4)dx_4\right]\right.\right.$$
$$\cdots \times \left.\left.\left[\int \psi_{25}(x_2, x_5)\psi_5(x_5)dx_5\right]dx_2\right]dx_1\right]$$

which leads to a much more efficient calculation since for each integral independent variables are removed, and thus, the total number of summations and multiplications is reduced. Interestingly, we can express the posterior marginal distribution as product

$$p(x_3) \propto \psi_3(x_3)m_{63}(x_3)m_{73}(x_3)m_{13}(x_3) \tag{2.23}$$

where we substituted the integrals by functions $m_{ij}(x_j)$. This decomposition is the major concept of message passing algorithms where the functions $m_{ij}(x_j)$ are regarded as messages passed through the graph. The messages $m_{63}(x_3)$ and $m_{73}(x_3)$ are sent from node 6 and 7; they are directly calculated by summing over x_6 and x_7

$$m_{63}(x_3) \propto \int \psi_{36}(x_3, x_6)\psi_6(x_6)dx_6$$

$$m_{73}(x_3) \propto \int \psi_{37}(x_3, x_7)\psi_7(x_7)dx_7$$

The message from node 1 also contains information from the remaining nodes of the graph and involves the message $m_{21}(x_1)$ from node 2 to 1

$$m_{13}(x_3) \propto \int \psi_{13}(x_1, x_3)\psi_1(x_1)m_{21}(x_1)dx_1$$

As indicated by arrows in Fig. 2.4 node 1 receives information from node 4 and 5 via node 2. Therefore, the message node 1 receives from node 2

$$m_{21}(x_1) \propto \int \psi_{12}(x_1, x_2)\psi_2(x_2)m_{42}(x_2)m_{52}(x_2)dx_2$$

contains also information that was passed from nodes 4 and 5 to node 2

$$m_{42}(x_2) \propto \int \psi_{24}(x_2, x_4)\psi_4(x_4)dx_4$$

$$m_{52}(x_2) \propto \int \psi_{25}(x_2, x_5)\psi_5(x_5)dx_5$$

This simple example shows how the posterior distribution can be calculated by propagating belief messages through the graph.

2.3.1 Belief Propagation

We will now consider the more general case, where we have an arbitrary pairwise MRF with a tree–structured graph $\mathcal{G} = (\mathcal{V}, \mathcal{E})$. In this case, *belief propagation* (BP) introduced by Pearl [179] and Lauritzen [120] describes an efficient message-passing algorithm for statistical inference. We can calculate the posterior marginal distribution, also called *belief*, of variable x_i by calculating the product of all incoming messages with the local observation $\psi(x_i, y_i)$

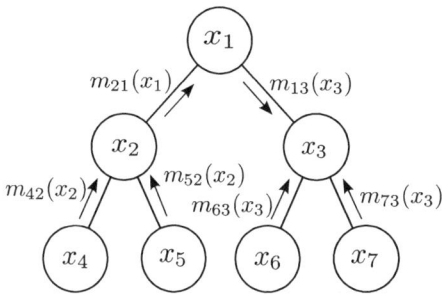

Fig. 2.4 Message passing example in a hierarchical structure: The belief of node 3 can be efficiently calculated by passing belief messages through the graph. The product of the incoming messages of node 3 gives the belief $p(x_3) \propto \psi_3(x_3)m_{63}(x_3)m_{73}(x_3)m_{13}(x_3)$.

$$b_i(x_i) = p(x_i|\boldsymbol{y}) \propto \psi_i(x_i, y_i) \prod_{j \in \Upsilon(i)} m_{ji}(x_i) \qquad (2.24)$$

This is a generalization of eq. 2.23. The message $m_{ji}(x_i)$ contains the "influence" the neighboring variable x_j has on x_i. It can be calculated in three steps. First, the *partial belief estimate* $b_{j\backslash i}(x_j)$ of node j is calculated similar to eq. 2.24 by combining all incoming messages except the one from node i with the local potential. Then, the influence of x_j on x_i is calculated via the pairwise potential $\psi_{ji}(x_i, x_j)$. Finally, x_j is marginalized out, so that $m_{ji}(x_i)$ is independent of x_j. Thus, the message $m_{ji}(x_i)$ from node j to node i can be written as

$$m_{ji}(x_i) \propto \int_{\mathcal{X}_j} \psi_{ji}(x_i, x_j)\psi_j(x_j, y_j) \prod_{k \in \Upsilon(j)\backslash i} m_{kj}(x_j)dx_j \qquad (2.25)$$

For a more detailed derivation of BP with a similar notation see for example [226]. If we, as before, consider the discrete case, where the variables can take one of K discrete values, the messages as well as the posterior marginal distributions are K–dimensional vectors. For a graph with N nodes the computation of all marginals then requires NK^2 operations since one message update $m_{ji}(x_i)$ needs K^2 operations. Compared to the K^N operations of the brute force implementation BP is able to significantly reduce the complexity.

Another possibility to further reduce the computational costs arises when pairwise potential functions $\psi_{ji}(x_i, x_j)$ only depend on the difference between its arguments $\psi_{ji}(x_j, x_i) = \psi_{ji}(x_j - x_i)$. In the discrete case the message update of eq. 2.25 then becomes

$$m_{ji}(x_i) \propto \sum_{x_j} \psi_{ji}(x_j - x_i)\psi_j(x_j, y_j) \prod_{k \in \Upsilon(j)\backslash i} m_{kj}(x_j) \qquad (2.26)$$

This computation can be seen as a discrete convolution

$$m_{ji}(x_i) = \psi_{ji}(x_j, x_i) * \prod_{k \in \Upsilon(j) \backslash i} m_{kj}(x_j) \tag{2.27}$$

where the convolution operator is defined as $(f * g)(x_i) \doteq \sum_{x_j} f(x_j - x_i)g(x_j)$. Thus especially if ψ_{ji} and the corresponding convolution kernel becomes large, it makes sense to replace the convolution by a multiplication in the Fourier domain.

2.3.2 Nonparametric Belief Propagation

Let us now consider the BP algorithm presented in Sec. 2.3.1 in the case where the random variables are represented by nonparametric distributions. The nonparametric formulation of the BP algorithm was introduced by Sudderth et al. [227] and Isard [100]. It is a generalization of the particle filter approach and assumes that the belief $b_i(x_i)$ as well as the messages $m_{ji}(x_i)$ are represented by sets of L discrete samples $\{(x_i^{(l)}, w_i^{(l)})\}_{l=1}^{L}$. In this case the posterior marginal distribution can not be calculated as in eq. 2.24. The problem is that the particles of the different messages will be distinct with probability one since they take values in a continues space and are sampled from independent proposal distributions. Applying eq. 2.24 and calculating the product would thus result in a distribution that is completely zero. Therefore, the kernel density estimate presented in Sec. 2.2 is used to interpolate between these samples. Gaussian kernels are used since they are smooth and strictly positive. Furthermore, they have the facilitating property that the product of d Gaussians $\mathcal{N}(x; \mu_1, \Lambda_1), ..., \mathcal{N}(x; \mu_d, \Lambda_d)$ is Gaussian $\mathcal{N}(x; \bar{\mu}, \bar{\Lambda})$ itself and can be calculated as

$$\bar{\Lambda}^{-1} = \sum_{i=1}^{d} \Lambda_i^{-1} \qquad \bar{\Lambda}^{-1}\bar{\mu} = \sum_{i=1}^{d} \Lambda_i^{-1}\mu_i \tag{2.28}$$

where $\bar{\mu}$ is the mean and $\bar{\Lambda}$ is the covariance of the resulting product distribution. Commonly weights are associated with the components of a Gaussian mixture model. The weight of the resulting Gaussian can be calculated as

$$\bar{w} \propto \frac{\prod_{i=1}^{d} w_i \mathcal{N}(x; \mu_i, \Lambda_i)}{\mathcal{N}(x; \bar{\mu}, \bar{\Lambda})} \tag{2.29}$$

with an arbitrary value for x (but $x = \bar{\mu}$ may be reasonable [226]). By convolution of the raw particle set with the Gaussian kernel the belief is constructed as

$$b_i(x_i) = \sum_{n=1}^{N} w_i^{(n)} \mathcal{N}\left(x_i; x_i^{(n)}, \Lambda_i\right) \tag{2.30}$$

These estimates are also used in regularized particle filters to avoid the degeneracy problem present in the standard particle filter approach [161, 7, 226]. The degeneracy problem describes the situation where all but one particle will have negligible weight. Equally to the belief, the messages are also approximated nonparametrically using Gaussian kernels. One problem that arises when approximating the belief is that the product of all incoming messages and the observation potential has to be calculated. Assuming that node x_i has d neighbors and that the observation potential is also represented by a mixture of Gaussians, then the product of $d+1$ Gaussian mixtures has to be calculated. If each mixture has L Gaussians then the resulting mixture will be comprised of Ld Gaussians. In order to handle this exponential growth several solutions have been proposed. One solution is to simplify the distribution by reducing the number of Gaussians. This solution can of course just be applied if the distribution has a simple form. Ihler et al. proposed more general solutions where the product mixture is approximated via a collection of independent samples [98]. The message update can be performed in two stages. First, L independent samples $\tilde{x}_j^{(l)}$ are drawn from the partial belief estimate

$$\tilde{x}_j^{(l)} \sim \frac{1}{Z_j} \varphi_{ij}(x_j) \psi_j(x_j, y) \prod_{k \in \mathcal{P}(j)\backslash i} m_{kj}(x_j) \qquad (2.31)$$

where the *marginal influence* function $\varphi_{ij}(x_j) = \int_{\mathcal{X}_i} \psi_{ij}(x_i, x_j) dx_i$ is used to reweight the samples. It captures the influence of $\psi_{ij}(x_i, x_j)$ on x_j. In the case where $\psi_{ij}(x_i, x_j) = \psi_{ij}(x_i - x_j)$ depends only on the difference between neighboring variables, as in this monograph, the marginal influence is constant and may be ignored. Then, the auxiliary particles $\tilde{x}_j^{(l)}$ are propagated to node i via the pairwise clique potential

$$\tilde{x}_{ji}^{(l)} \sim \frac{1}{Z_j^l} \psi_{ij}(x_i, x_j = \tilde{x}_j^{(l)}) \qquad Z_j^l = \int_{\mathcal{X}_i} \psi_{ij}(x_i, x_j = \tilde{x}_j^{(l)}) dx_i \qquad (2.32)$$

The message is therefore defined by the samples $\tilde{x}_{ji}^{(l)} \in \mathcal{X}_i$ drawn from the conditional density $\psi_j(x_i|x_j)$ of each fixed auxiliary particle $\tilde{x}_j^{(l)}$. Finally, the N particles $\tilde{x}_j^{(l)}$ are used to construct a nonparametric estimate of the target message $m_{ij}(x_i)$. For that, we will apply the "rule of thumb" in this monograph to determine the bandwidth λ_{ji}. For an overview of other (automatic) bandwidth selection methods consult [215].

2.3.3 Scheduling the Message Passing

Up to now, we have seen how a message from one node to another can be calculated. Still undiscussed is the question, in which order the messages have to be calculated. This task is solved by creating a schedule by which the messages are updated [179, 226]. Node i can only send a message to node j when node i has received messages from all of its other neighbors (except node j). Initially, this is just the case for the leaves. In our hierarchical model, an appropriate schedule starts therefore with a bottom-up sweep propagating the evidence of the leaves to the root node. In a top-down sweep, the information of the root is then propagated back to the leaves in the reverse direction. After these two stages the exact posterior marginal distribution of each node can be calculated. Another schedule, which is especially suitable for distributed implementations, is a parallel version of the BP algorithm. Instead of sequentially calculating the messages as before, the schedule proceeds in iterations, in which all nodes use their received messages from the previous iteration to recompute the new messages [226].

Chapter 3
Hierarchical Graphical Models

The goal of the proposed hierarchical graphical models is to recognize various instances of different object classes in images, image sequences or other scene representations like e.g. occupancy grid maps. The term "object" in this context is used as a general term representing visual objects, visual parts, visual features, visual primitives, but also activities, actions or motion primitives. In the following we will regard two different kinds of hierarchies: A compositional hierarchy and a similarity hierarchy. In compositional hierarchies the structure of a parent node is defined by its children, where edges define the spatial or spatiotemporal relation between the parent and the children nodes. In this manner complex high-level nodes can be recursively defined based on simple low-level features. Similarity hierarchies, on the other hand, describe similarities among objects, and among parts. In this work, they will be combined with a coarse-to-fine search by means of scale space representation. They are used to increase the robustness of the representation as well as the overall runtime performance.

Compositional hierarchies have several advantages over other representations. Their compositional structure allows us to build flexible and sparse object models. The flexibility is guaranteed on the one hand by the edges that are able to represent arbitrary spatial relations and on the other hand by the hierarchical structure that represents complex objects by decomposing them into simple parts. This also leads to good generalization properties. The local features and part nodes of the hierarchy adapt to variations of the local observations while the high-level nodes maintain the overall spatial arrangement of the parts. Using a finite data set of instances during training the model thus can perform accurately on new, unseen instances. Furthermore, sparsity allows to reduce the number of relevant low-level features. Thus, instead of representing an object by image patches that include a large amount of color information, the hierarchical structure and its local feature relations allow to focus on a feature subset like e.g. edges. This reduces the influence of clutter but also reduces the processing time since the complexity is kept tractable. Another important property of hierarchies is that parts and features can be shared among different objects and configurations. This reduces the storage demands since parts have to be stored just once but can be used multiple times. Simultaneously, the computa-

© Springer International Publishing Switzerland 2015   21
J. Spehr, *On Hierarchical Models for Visual Recognition & Learning of Objects, Scenes, & Activities,*
Studies in Systems, Decision and Control 11, DOI: 10.1007/978-3-319-11325-8_3

tional effort decreases during the bottom-up inference calculations since the partial belief estimates have to be calculated just once and can be sent to all parent nodes. Learning and especially structure learning is another field where hierarchies show significant advantages. Structure learning is in general a very challenging problem. For arbitrary structures the problem is even unsolved. Compositional hierarchies, however, offer some interesting structural properties which simplify the learning process essentially. Especially during online learning, sharing of parts allows to learn and express new instances by means of already learned parts. This reduces the complexity of structure learning and makes it tractable for the applications in this monograph.

We will now give a brief outline of the following sections. First, we will formally define our hierarchies. The related work is divided into different models and also into related principles. After a simple example of a compositional hierarchy we describe the different information sources of a node, and show how the spatial (also spatiotemporal) dependencies are incorporated into the hierarchy. Then, we will introduce our new compositional and similarity hierarchies including important properties such as scale, orientation and view invariance and coarse-to-fine inference methods. We also propose new efficient inference approaches, which will be presented at the end of this chapter.

3.1 Hierarchical Graphical Models

Based on the definitions introduced in the previous chapter, we can now formally define the hierarchical structure, which will be extensively used in the following chapters. The hierarchy is represented by an undirected tree-structured graph $\mathcal{G} = (\mathcal{V}, \mathcal{E})$. More specifically, the graph is a *rooted tree* where exactly one node is appointed as *root*. Each node in the tree is assigned to a level ℓ. The root node is always assigned to the highest level in the tree. In relation to the root node the edges have two directions: bottom-up (towards the root) and top-down (away from the root). Similar to directed graphs, we define the set of parents $\Gamma(j)$ as $\Gamma(j) = \{i \in \mathcal{V}|(i,j) \in \mathcal{E} \wedge \ell_i > \ell_j\}$. In a tree structured graph, every node except the root node has an unique parent. The set of all children $\Xi(j)$ is given by $\Xi(j) = \{k \in \mathcal{V}|(j,k) \in \mathcal{E} \wedge \ell_i < \ell_j\}$. A node without children is called a *leaf*. Please note, that we do not restrict the parents to be on the next higher level or the children to be on the next lower level. Without this restriction the hierarchies get more flexible and can more accurately model the real world as we will see in the next section. Throughout the following chapters we will use pairwise Markov random fields to model the statistical dependencies between the nodes in the hierarchy. We have seen in Sec. 2.1.2 that the marginal distribution of a node can be written as the product of pairwise potential functions according to eq. 2.8. The example tree shown in Fig. 3.1(a) is a hierarchy with three levels. Although we are using undirected graphs the relation to directed graphs is obvious (a directed version of the graph can be seen in Fig. 3.1(b)). As described in Sec. 2.1.3, a directed graph can be converted to an undirected graph. For that, we have to replace the undirected edges with directed ones. The corresponding potential function $\psi_{cp}(x_c, x_p)$ between a parent node x_p and one of its children x_c equates to the conditional distribution $p(x_c|x_p)$. Since the nodes in the hierarchy have not more than one parent a moralization is not necessary. Just a prior distribution of the root node has to be added to the marginal distribution.

3.2 Related Work

In this section, we will give an overview of models and especially hierarchical models used in computer vision so far. Literature related to our applications will be separately summarized at the beginning of the following application oriented chapters. First, in Sec. 3.2.1 we will present an overview of appearance-only models, also called bag-of-words models, which ignore the geometrical properties between features. These models are less related to our hierarchy but are of large interest in the computer vision community due to their simplicity. In Sec. 3.2.2 we summarize part-based models, which consider objects as a composition of parts and model the geometrical structure between them. Constellation models are a special kind of statistical part-based models and are briefly reviewed in Sec. 3.2.3. Most related to our approach are compositional hierarchies presented in Sec. 3.2.4. We additionally review approaches touching important properties of our compositional hierarchy. First, we give a summary of related work concerning the sharing of features and parts among different hierarchies in Sec. 3.2.5 and, subsequently, review hierarchies, which are used in the context of coarse-to-fine search in Sec. 3.2.6. Finally, we conclude with a brief review of biologically inspired models in Sec. 3.2.7.

3.2.1 Appearance-Only Models

One major class of object recognition models is the appearance-only representation where the location of image features is disregarded [204]. It was originally inspired by text retrieval techniques [9]. The text is parsed word by word and the frequency of occurrence of the words in the document is determined. The whole document is then represented by a vector, where each component stores the frequency of a specific word. This representation is also called bag-of-words model. The idea can be transferred to computer vision tasks. In this case the words are image features, and the document is the image. They are therefore also referred to as bag-of-features approaches. Each approach can be divided into an offline training step, in which a vocabulary and an appropriate classifier is learned, and an online recognition step. During training, a feature detector is applied to the training images and feature descriptors of the local regions around the features are extracted. Then, a clustering algorithm is applied to the descriptors, where each cluster center corresponds to a visual word (also called "texton", "codeword", "patch" or "keypoint") of the final vocabulary. This vocabulary is then used to describe the content of an image. For that, the detected features and their descriptors are assigned to the nearest word of

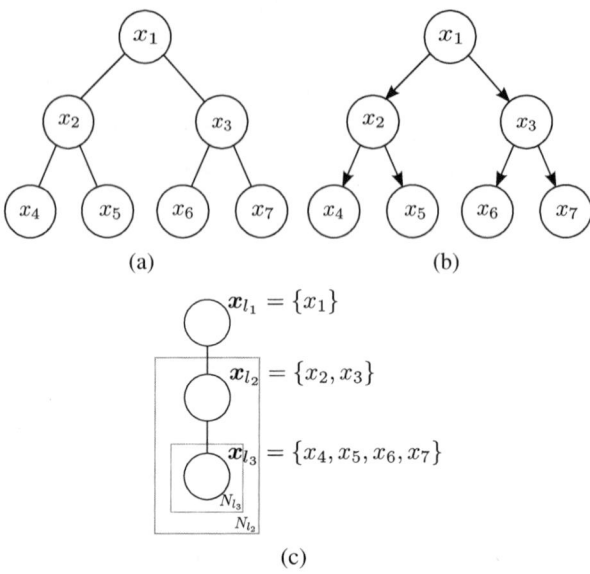

Fig. 3.1 Hierarchical Graphical models: (a) Hierarchical Markov random field. (b) Hierarchical Bayesian network. (c) Plate notation.

the vocabulary, and the frequency of each visual word is stored in a histogram vector. The whole image content is represented by this histogram and is used as the feature input vector during classification.

Many similar appearance-only models have been published during the last decade. Although they disregard any location and geometrical information, they are able to achieve excellent classification results on standard recognition datasets. One of the first bag-of-features approaches was developed by Sivic and Zisserman [216] and adapted to the problem of object matching. They described the object by a set of viewpoint invariant region descriptors. Using k-means clustering, the descriptors were assigned to clusters which form the visual "words". The vector of word frequencies was then used for object and frame retrieval in video sequences. Dorkó and Schmid [44] applied the idea to object class recognition. They determined local "Scale Invariant Feature Transform" (SIFT) descriptors [134] in order to characterize the object class appearance. The features were then clustered using the Expectation-Maximization (EM) algorithm. Then, ranking-based feature selection was used to determine the most discriminative parts. Csurka et al. [35] used bag-of-keypoints for visual categorization. They detected Harris affine points and represented the local region by the SIFT descriptor. After vector quantization, a bag-of-keypoints was constructed which counts the number of patches assigned to each cluster. A multi-class classifier was finally applied to the bag-of-keypoints in order to determine which category to assign to the image. Winn et al. [249] extended this work by introducing dense feature extraction and an automatical learning of the optimal keypoints and dictionary size.

Another model that uses the bag-of-words representation is the topic model (e.g. Latent Dirichlet Allocation [17]). The main idea of a topic model is that documents are mixtures of hidden topics and that these topics are probability distributions over words. Since topic models are generative, they can be used to make new documents. For that, we first have to choose a distribution over topics. Then, we use this distribution to choose a topic at random. And finally words are drawn from that topic. Fei-Fei and Perona [129] applied this idea successfully to the learning and recognition of natural scene categories. Here, the topics were themes that represent specific image contents. For a "mountain" class, a topic could e.g. be a "rock" theme. This theme has a specific distribution over typical visual words (like e.g. slanted lines). Sudderth et al. [228] proposed a similar model that has a distribution over parts for each object category, and for each part a distribution over expected appearances and positions.

3.2.2 Part-Based Models

Part-based models are considering objects as a composition of parts and are modeling the geometrical structure between them. This idea dates back to Fischler and Eschlager [68] who introduced the "parts and structure" model that separates an object into its parts and uses relation functions to model the spatial dependencies between them. An example is a face which is defined by its parts (hair, eyes, mouth,...) and where the pairwise relations are represented by springs. This model inspired many other works that differ in the kind of object and their parts, the low-level description, the statistical framework, and the computer vision task [57, 214, 261, 117]. Wahl and Biland [241, 14] described complex polyhedral scenes by their constituent components and decomposed these components further into object primitives, such as tetrahedra, prisms and parallelepipeds. The whole interpretation was performed in the Hough space. Engelbrecht and Wahl [47] applied this idea to object recognition which was formulated as a graph matching problem. They used an attributed subgraph isomorphism algorithm in order to compare the detected graph representation to similar representations of CAD models. Leibe et al. [125, 127, 126] proposed a part-based model where the parts are represented by appearance patches. These patches were extracted around Harris interest points from a set of training images and were clustered to form a codebook. The spatial distribution of the codewords was learned and used to define the object. Finally, in order to detect the object a probabilistic extension of the generalized Hough transform was used as a voting scheme. Opelt et al. [178] extended this idea and added shape information. They used boundary fragments and appearance patches to represent the object and a Hough voting scheme to detect the object. Similar to Opelt, Ferrari et al. [62, 61] were using local contour features. They grouped connected and approximately straight contour segments, which they called k-adjacent segments (kAS) and used them to formulate a codebook. The object was represented by a histogram which counts the number of the different kAS types within a detector window. In

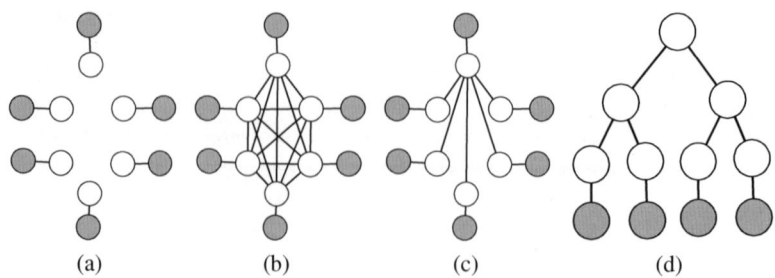

Fig. 3.2 Different kinds of graphical models commonly used in computer vision (modified from [204]). a) Bag-of-words [216, 44, 35, 249, 17]. b) Fully connected [58]. c) Star model [59]. d) Three layer hierarchy [19].

order to encode spatial information into the model, they also proposed to subdivide the window into a set of tiles and calculate a kAS histogram separately for each tile.

3.2.3 Constellation Models

Burl et al. [27] introduced a probabilistic approach which models objects as random constellations of parts. Due to the probabilistic formulation of the part relations it is able to model shape variations, image clutter, occlusions and detector errors. Later, Weber et al. [245, 246] introduced a maximum likelihood weakly supervised learning algorithm for the constellation model, which learns class models from unlabeled and unsegmented cluttered scenes. Fergus et al. [59, 60] extended this work by modeling the variability of appearance, which is simultaneously learned with the shape. Furthermore, they used an interest operator to detect regions and their scale, and added scale invariance to the representation.

There are three different types of constellation models. In the *fully connected* model [58], each part of the constellation is connected to all other parts. This provides the most general description, but simultaneously makes the learning difficult due to the high number of model parameters. The *star* model [59] solves this problem by a simplified configuration model. It is a tree of depth one, where a landmark part L is defined as the root node. The advantage is that all parts, except the landmark, are conditionally independent of each other. Thus, according to the naive Bayes assumption, the joint distribution factorizes as

$$p(\boldsymbol{X}|\boldsymbol{S}, \boldsymbol{h}, \theta) = p(x_L|h_L) \prod_{j \neq L} p(\boldsymbol{x}_j|\boldsymbol{x}_L, s_L, h_j, \theta_j) \qquad (3.1)$$

where x_j is the position of part j, $\boldsymbol{X}$, $\boldsymbol{S}$ are the locations and scales of image features, $\boldsymbol{h}$ is an indexing variable, which allocates a particular feature to a part, and θ are the model parameters. Fergus et al. applied this model successfully to a wide variety of categories for classification as well as for localization of the object within the image. Felzenszwalb and Huttenlocher proposed a similar tree model [57] and further improved the recognition performance by means of distance-transforms.

The *k-fan* model [34] is defined by a set of k reference parts, which have the same purpose as the landmark part in the star model. Indeed, if $k = 1$ the model corresponds to the star model and if k equals the number of parts the model is fully connected. Since a k-fan model can be seen as a collection of cliques of size $k + 1$ one can apply eq. 2.4 to calculate the joint distribution. For their investigated classes of objects, Crandall et al. showed that a small amount of spatial structure can provide similar recognition performance as more powerful models.

3.2.4 Compositional Hierarchies

The idea of compositional hierarchies has a long history. In early works like e.g. Marr and Nishihara 1978 [142] 3d objects were represented as a hierarchy of simple primitives. The parts and primitives were often compositions of predefined simple features like corners or inflection points, and of compound features like cranks or bumps [53].

More recently, Geman et al. [80] proposed a mathematical formulation of compositionality which is inspired by the Minimum Description Length Principle and promotes a recursive grouping of constituents. The approach was used in [104] for reading license plates and achieves an accuracy of 98%. Ommer and Buhmann [175, 176, 177] proposed a hierarchical model that uses localized histograms to define the features on each layer. The local histograms [175] were local edge and color histograms of subpatches and delivered a low dimensional representation of an image patch. They restricted the hierarchy to a single layer of compositions since in their experiments this had proven to be sufficient [176]. Epshtein and Ullman [48] proposed a hierarchy that is learned in a top-down manner. The approach starts at the top-level fragments and repeatedly breaks down the fragments into their own optimal components. An optimal component is thereby a sub-feature that often appears in regions of the parent feature and seldom elsewhere. In experiments, they showed that the hierarchical structure significantly improves the performance compared to a holistic (non-hierarchical) approach. Later they extended their work to semantic hierarchies [49]. Bouchard and Triggs [19] introduced an extension of the constellation model [60]. They used a three layer hierarchy, with the object at the top level, the parts at the center level and the local image features at the bottom level. While the constellation model typically encodes 6 or at most 7 features, which correspond in this model to the parts, the three layer hierarchy is capable of handling hundreds of local features. This makes the representation suitable to very basic feature detectors like Harris keypoints with SIFT descriptors. Todorovic and Ahuja [233] made use of a recursive image segmentation process and represented an image as a hierarchy of segmented parts. During learning they determined a segmentation tree for each image of the training set and matched the extracted trees among each other in order to find maximally matching subtrees. These subtrees were taken as instances of the target category and were fused into a tree-union forming the canonical model of a visual category. Another class of hierarchy and learning approaches was regarded by Scalzo and Piater [205, 204], L. Zhu et al. [268], Fidler and Leonardis [66, 63]. These approaches have in common, that they learn the hierarchy in a bottom-up manner. They start at the bottom and try to find correlated features that often occur within a neighborhood. These feature are then combined and define a new compound feature. The approaches use different kinds of low-level features: SIFT [205], invariant triplet vector [267, 268] and Gabor wavelets [66] and describe the objects with a hierarchy of 3-5 layers.

Detry et al. [41] used a hierarchy to directly model 3d objects. Their probabilistic framework is similar to ours, except that they were using random variables and pairwise potential functions in 3d space. As a consequence, the variables and the potentials have to be defined over the Special Euclidean group $SE(3) = \mathbb{R}^3 \times SO(3)$ with the Special Orthogonal group $SO(3)$ for the 3d orientation. Furthermore, the Dimroth-Watson distribution, which corresponds to a Gaussian-like distribution on $SO(3)$, has to be used in order to handle the double cover of $SO(3)$ by quaternions. These additional modeling costs lead to an increase of the computational effort.

Another hierarchical model for object representation is the AND/OR graph [39]. Zhu and Mumford [269] represented objects by a hierarchical stochastic grammar that is embodied in a simple AND/OR graph representation. The AND-nodes are decomposed into a number of components and the OR-nodes point to alternative sub-configurations. Another AND/OR graph representation was proposed by Chen et al. [30]. Their model is capable of describing the different configurations of deformable articulated objects. The proposed inference algorithm combines a bottom-up process for proposing configurations together with a top-down process for refining and validating these proposals.

3.2.5 Feature and Part Sharing

Feature and part sharing is a common topic in computer vision and of special interest for compositional hierarchies. Their characteristics considering sharing and reusability enables a compact representation of multiple objects, lead to better generalizations and also improve the efficiency of inference and learning [234]. The importance of reusability was already mentioned in early works of Biederman [13]. In the "recognition-by-components" framework objects are decomposed into parts, also called geons, which are based on simple 3d shapes like cylinders or cones. Biederman suggested that a small set of less than 36 geons is sufficient to constitute a large ensemble of real world shapes and objects. Krempp et al. [114] investigated the sequential learning of reusable parts and showed that the number of distinct parts in the system grows slower than the number of classes. More recently, sharing has been integrated in many common models. In general, all approaches using a codebook of reusable low-level visual words can be seen as 'sharing' approaches [216, 175, 44, 134, 35, 129, 228] . Thomas et al. [231] used activation links for multi-view object class detection. The activation links are defined between single-view models and are used to share votes among each other during recognition. A similar approach was proposed by Savarese and Fei-Fei [202]. They linked together diagnostic parts of the objects from different viewing points. One disadvantage of both approaches is that the structure is restricted to two-layers. Torralba el al. [234] proposed feature sharing in a boosting framework for detecting a large number of different classes. The final classification process ran faster and required less training samples. Fidler and Leonardis [66, 63] used hierarchical structures with up to five levels and part sharing. They allowed simple primitives to be shared by multiple parent nodes, but they did not share information between multiple-views. Mikolajczyk et al. [148] shared appearance clusters built from edge-based features among several object classes, but did not share information between multiple views. Zhu et al. [266] proposed part and appearance sharing between different object classes and views in a compositional model, but they did not consider different scales and their model is restricted to five layers.

3.2.6 Coarse-to-Fine Hierarchies

In computer vision, decision trees are often used to describe hierarchies which are suitable for classification tasks but not for generative object representation [160]. They are connected to search trees, which refer more to the underlying data structure. The aim of a decision tree as a classifier is to assign some sort of output, represented by the leaves, to a given input, represented by the root. A specific output is selected by a sequential decision making process starting at the root and traversing depth-first to the leaves. At each decision node a branch is taken depending on the chosen decision. One commonly used decision tree is a binary decision tree where each node splits into two branches. The complexity of the classification is in general logarithmic since the depth of a binary tree with n leaves is at least $\log n$. One well-known binary search tree is the kd-tree, which is usually used as an efficient data structure for nearest-neighbor search problems. It recursively partitions the search space into half-spaces until finally the leaves just contain single points.

Similar concepts have been applied in the context of point matching and image retrieval. The main idea of these applications is, first, to extract features from the query image or query point, and, second, to use a decision tree to efficiently index an image of the database or a keypoint. Obdrzálek and Matas [171] used a binary decision tree to index keypoints for image retrieval. A similar approach was proposed by Lepetit et al. [128], who were using multiple randomized decision trees to efficiently index keypoints. Maree et al. [141] used randomly extracted subwindows as input for an ensemble method of decision trees, which combines the predictions of several models. Nistér and Stewénius [169] performed image retrieval by means of a vocabulary tree that hierarchically quantizes the descriptors detected in the images. Philbin et al. [181] used a similar approach but incorporated spatial information by means of a RANSAC and re-ranking technique.

Other methods that can be interpreted as decision trees are coarse-to-fine approaches, where imprecise initial solutions are iteratively refined. Similar to decision trees, they start with a coarse solution, represented by the root node of the tree, and refine their solution step by step by means of some local decisions until the fine solution is reached at the leaf. Gavrila [76, 75] proposed a hierarchical template matching, where a large number of objects (templates) is efficiently matched with an image using distance transform. Instead of matching the template separately with the image, the search is incorporated into a hierarchy. The hierarchy is built by first grouping similar templates together and representing them by a prototype, and then recursively repeating the grouping at various levels. During matching, the algorithm starts at the root node and traverses the tree in depth-first fashion. Lin et al. [130] extended this hierarchical template matching by decomposing the global shape models into parts. They applied the approach to the detection of human shapes. The algorithm starts by matching the head-torso shape first, and then hierarchically match the upper and lower leg template models. Fleuret and Geman [71] proposed a coarse-to-fine sequential binary testing of the presence or absence of loose spatial arrangements. Here, the order of the sequential binary tests is chosen to minimize the overall computation. A combination of compositional hierarchies and scale hierarchies was proposed by Ettinger [54]. The sub-part hierarchy was used for efficient indexing and relative parameterization among sub-parts, while the scale hierarchy was used for a coarse-to-fine recognition scheme. Fidler et al. [65] proposed a taxonomy of constellation models similar to the hierarchical template matching of Gavrila [76]. They cascaded the recognition process from coarse-to-fine resolution and built the taxonomy on the highest level of a compositional hierarchy.

3.2.7 Biologically Inspired Models

Biologically inspired models are trying to explain the visual processing in the human cortex. They are based on some widely accepted facts about the ventral stream, which is associated with object recognition and form representation in the visual cortex [95, 96, 210, 211]. These widely accepted facts are:

- Visual processing is hierarchical, aiming to build invariance to position and scale first and then to viewpoint and other transformations.
- Along the hierarchy, the receptive fields of the neurons as well as the complexity of their optimal stimuli increases.
- The initial processing of information is feedforward (for immediate recognition tasks).
- Plasticity and learning probably occurs at all stages and certainly at the level of inferotemporal cortex and prefrontal cortex, the top-most layers of the hierarchy.

One approach using these facts is the HMAX approach [192] that is based on Fukushima's neocognitron model [73]. The layers of the model consist of alternating computational units: simple S units and complex C units. The S units perform linear operations and can be seen as template matching units. The C units perform non-linear pooling operations where their inputs are combined by a maximum (MAX) function. While the S units increase selectivity, the C units increase invariance. The visual processing starts with the first S_1 layer consisting of simple cell-like receptive fields that are modeled by a set of first Gaussians derivative filters computed at different scales and orientations. The responses of the first layer are combined by the cells of the second C_1 layer by performing a max operation to the filter responses of the same orientation, but different scales and positions. The S_2 layer combines the responses of the C_1 layer to form more complex feature detectors. The global maxima over all scales and positions are then taken by the C_2 layer achieving invariance regarding shift and scale. In the HMAX model, the C_2 units are finally fed into "view-tuned units". Serre et al. [211] extended this model. While the original model used a simple static dictionary of manually defined features, they learned a vocabulary of visual features from images and applied it to the recognition of real-world object categories. Mutch and Lowe [162] refined this approach using simple versions of sparsification and lateral inhibition and were able to achieve state-of-the-art recognition performance. Masquelier and Thorpe [144] also exploited the four-layer hierarchy (S_1–C_1–S_2–C_2) but proposed to use spiking neurons and operate in the temporal domain, where the processed information is defined by the time between the first spike and the stimulus onset. Furthermore, they proposed an unsupervised learning approach based on Hebbian rules to build the hierarchy.

Other biologically inspired models are convolutional neural networks [121, 122]. They are multi-layer networks with alternating convolution and subsampling layers, which are e.g. used in [122] for document recognition and in [209] for traffic sign recognition.

3.2.8 Comparison of the Related Work with this Monograph

The hierarchical models discussed so far cover a wide range of concepts and applications. However, the hierarchical structure is very restricted. Typically, the hierarchies have up to 5 levels and nodes have up to four children, where the children have to lie on the next lower level. Unfortunately, these restrictions lead to inefficient representations, where the maximal reusability of parts and primitives can not be reached. In our hierarchical model the level of an object is directly determined by the number of associated low-level features. This guarantees an unified hierarchical framework with flexible inter layer dependencies leading to a compact and efficient representation. Different to previous models, we directly integrate different scales and orientations of an object and its parts into the hierarchy. Here, all scales and orientations of one specific object share the same building rule, which defines the decomposition. This simplifies the structure and also the learning since the decomposition has just to be learned once. Furthermore, this sharing guarantees that not just the objects but simultaneously all its parts are represented at all scales and orientations.

In order to further increase the performance of our representation we also propose to use similarity properties between the objects as well as the parts. In contrast to previous models, which establish similarity links directly between nodes or views [202, 64], we are proposing similarity hierarchies to reduce the number of links and to integrate a coarse-to-fine scheme into our model. We couple the levels of the similarity hierarchy with a scale space representation, allowing us (1) to efficiently generate hypotheses at a coarser scale using the composition hierarchy, (2) to refine them using the similarity hierarchy and (3) to evaluate them at a fine scale using the compositional hierarchy again. As our results will show, the combined compositional and similarity hierarchy outperforms standard compositional hierarchies especially in cluttered scenes.

We model our hierarchy using a Markov random field and refer to well-established methods like kernel density estimation [215], nonparametric belief propagation [227] or product estimation techniques [98]. This allows us to precisely formulate the hierarchical model, make use of efficient inference techniques, and, furthermore, propose improvements of the standard techniques: (1) combination of the bottom-up message passing with a top-down passing step, (2) bottom-up passing in several sequential sweeps, and (3) importance sampling for high-level observable nodes.

Another difference is the kind of dataset, we are using to learn our hierarchies. Rather than using common benchmark datasets supporting hundreds of instances for one specific object class, we are learning single instances, which are for example elements of a pose collection. Our learning is therefore an instance-based learning, where the reuse of parts avoids the linear growth with the number of instances.

Also novel is the use of compositional hierarchies in combination with sharing for human pose estimation, activity representation and scene understanding. We will highlight the contributions in these fields separately at the beginning of each application chapter.

3.3   A Simple Example

We begin by illustrating the compositional concept using the simple hierarchy shown in Fig. 3.3. The face is decomposed into its parts like forehead, the sides including the ears, the eyes, the chin, the nose and mouth. These parts can again be decomposed into smaller parts like hairline, a single eye, or a part of the chin. And, finally, these small parts are decomposed into visual primitives, which represent basic elements like edges or corners. The compositional hierarchy consists of two main components: (1) the low-level features and (2) the spatial relations between the features, parts, and primitives. The low-level features are representing the observations, and they are gathered directly from the input image. All higher levels can not be observed, they are hidden, and information has to be inferred from lower levels using the spatial relations. These spatial relations determine the relative positions of an object to its parts.

Since an object can be decomposed in various ways it can be represented by different hierarchies. This is an important property allowing us in this monograph to choose that hierarchy, which is attractive from a computational point of view. The idea of computational efficiency is to reduce redundant calculations, and thus calculate similar parts, like the left and right eye, just once.

3.4   Nodes in the Hierarchy and Their Information Sources

We will now discuss which information sources a node in the hierarchy can access. We will distinguish between four different types of information sources: evidence,

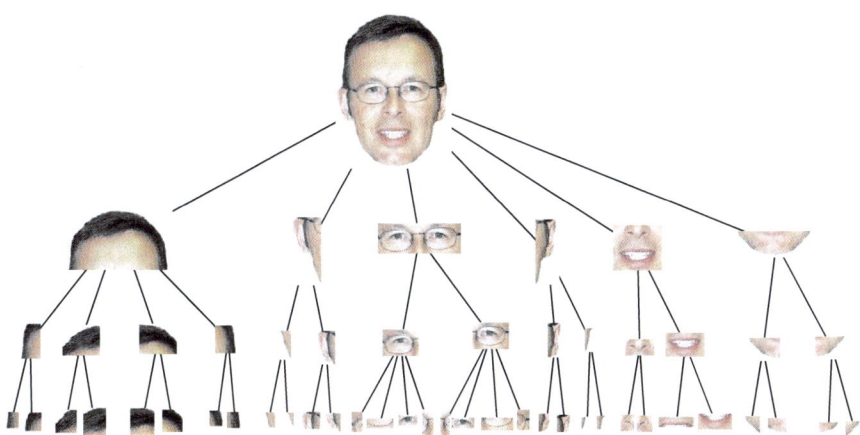

Fig. 3.3 Simple example of a hierarchical decomposition: The face is decomposed into its parts, these parts are again decomposed, and so on. The edges define the spatial relations between elements (face taken from Caltech-256 [83]).

composition, similarity and time (depicted in Fig. 3.4 with different colors). As described in Sec. 2.3.1, the node i can determine the posterior marginal distribution $b_i(x_i) = p(x_i|\boldsymbol{y})$ by combining the information of all neighbors

$$b_i(x_i) \propto p(x_i|\boldsymbol{y}_{obs})p(x_i|\boldsymbol{y}_{comp})p(x_i|\boldsymbol{y}_{sim})p(x_i|\boldsymbol{y}_{temp}) \tag{3.2}$$

The observation function $p(x_i|\boldsymbol{y}_{obs})$ is used to induce evidence into the hierarchy and can be divided into a local and global observation function

$$p(x_i|\boldsymbol{y}_{obs}) = \phi_i(x_i, y_i)p(x_i|\boldsymbol{y}_{\Theta(i)}) \tag{3.3}$$

where $\boldsymbol{y}_{\Theta(i)}$ denote all observations that reach x_i over the global contextual nodes $\Theta(i)$. The local observation function $\phi_i(x_i, y_i)$ provides local evidence gathered by local feature extraction and is represented by an edge to the observable node y_i (shaded in Fig. 3.4). On the other hand, the global observation function $p(x_i|\boldsymbol{y}_{\Theta(i)})$ provides contextual information (global evidence) and could be directly gathered using global features. The difference between local and global features will be further discussed in Sec. 3.5.1. In this monograph we will just use local features, so that $p(x_i|\boldsymbol{y}_{obs}) = \phi_i(x_i, y_i)$ is the relation between an observed node y_i representing the local evidence, and the hidden node x_i. Only the observable nodes are connected to local observations y_i, thus, $\phi_i(x_i, y_i)$ can be omitted for hidden nodes.

Information provided by the compositional hierarchy are regarded by $p(x_i|\boldsymbol{y}_{comp})$, which can be divided into information received from the parents and children

$$p(x_i|\boldsymbol{y}_{comp}) = p(x_i|\boldsymbol{y}_{\Gamma(i)})p(x_i|\boldsymbol{y}_{\Xi(i)}) \tag{3.4}$$

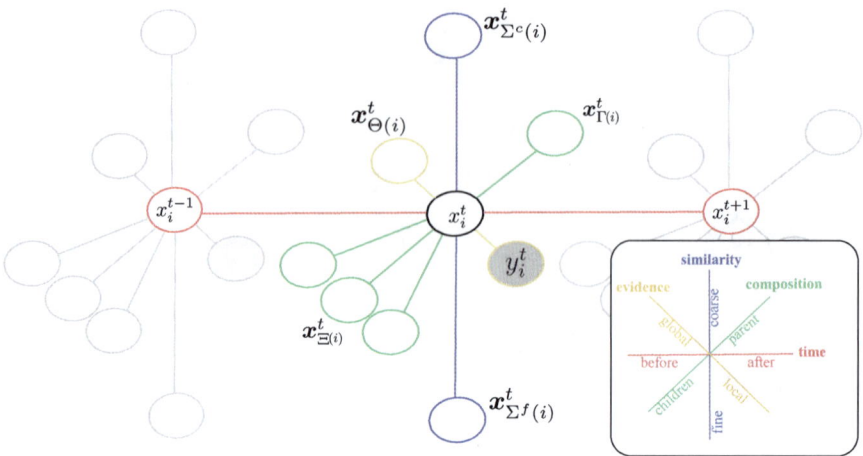

Fig. 3.4 Information sources of node x_i (evidence, composition, similarity and time)

Let y_Γ denote all observations that reach x_i over the parent nodes. Then, the distribution $p(x_i|y_\Gamma)$ represents the estimate that defines how likely the parent nodes think it is that node x_i will be in a corresponding state. Since the parent nodes are in general not directly observable, they gather information from the low-level feature nodes during the bottom-up message passing step and sent it to node x_i during the top-down step (see Sec. 3.5.3). The parent nodes capture the structural dependencies and guarantee the overall arrangement of the parts. The parent information can, hence, also be seen as a spatial context. It is defined as:

$$p(x_i|y_\Gamma) = \prod_{j \in \Gamma(i)} m_{ji}(x_i) \tag{3.5}$$

The children nodes $\Xi(i)$ connect node i directly or indirectly over the child's children to the low-level feature information. It thus captures the local evidence of the node. If y_Ξ denotes all observations that reach x_i over the children nodes then the local evidence of x_i is given by

$$p(x_i|y_\Xi) = \prod_{j \in \Xi(i)} m_{ji}(x_i) \tag{3.6}$$

In vertical direction, node x_i receives information provided by similar nodes $p(x_i|y_{sim})$. Here, we distinguish between similar nodes at a finer scale $\Sigma^f(i)$, the same scale $\Sigma^s(i)$ and at a coarser scale $\Sigma^c(i)$

$$p(x_i|y_{sim}) = p(x_i|y_{\Sigma^c(i)})p(x_i|y_{\Sigma^s(i)})p(x_i|y_{\Sigma^f(i)}) \tag{3.7}$$

The links are used to share information between equal and similar nodes. The information allows to speed up the calculation since the information can be shared

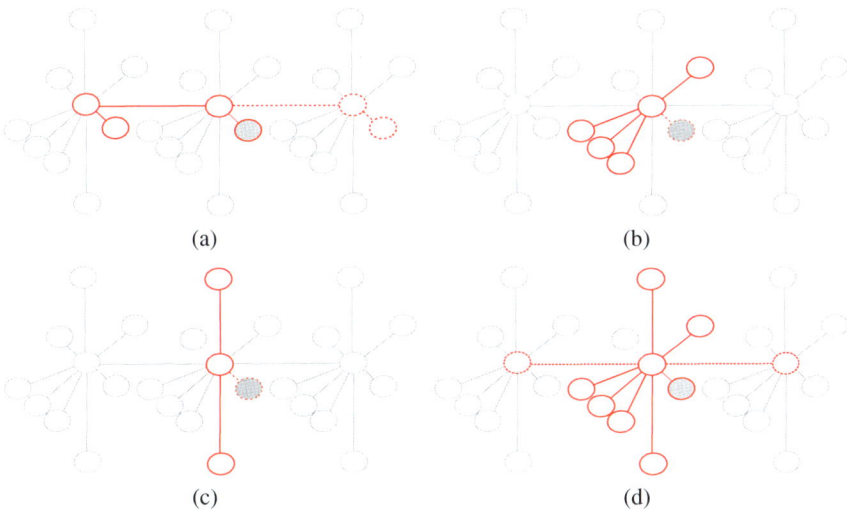

(a)   (b)

(c)   (d)

Fig. 3.5 Examples of information sources (red) used in the following approaches: a) Temporal tracking (e.g. Kalman- or particle filter). b) Compositional hierarchies (e.g. [205, 268, 66]. c) Similarity hierarchies (e.g. [76, 75]). d) Combined similarity and compositional hierarchy (this monograph).

between equal nodes and thus has to be calculated just once. Furthermore, the robustness can be improved since similar nodes should have similar spatial distributions. Each distribution summarizes the information gathered at the particular scale and they are defined as

$$p(x_i|\boldsymbol{y}_{\Sigma^c}) = \prod_{j \in \Sigma^c(i)} m_{ji}(x_i) \tag{3.8}$$

$$p(x_i|\boldsymbol{y}_{\Sigma^s}) = \prod_{j \in \Sigma^s(i)} m_{ji}(x_i) \tag{3.9}$$

$$p(x_i|\boldsymbol{y}_{\Sigma^f}) = \prod_{j \in \Sigma^f(i)} m_{ji}(x_i) \tag{3.10}$$

When we regard a dynamic system at discrete time steps, we also have information from the previous $1 : t - 1$ and the next $t + 1 : T$ time steps

$$p(x_i|\boldsymbol{y}_{temp}) = p(x_i|\boldsymbol{y}_i^{1:t-1})p(x_i|\boldsymbol{y}_i^{t+1:T}) \tag{3.11}$$

In tracking applications we are in general just interested in $p(x_i|\boldsymbol{y}_i^{1:t-1})$, which corresponds to the prediction of x_i based on the information of the previous time steps. The distribution $p(x_i|\boldsymbol{y}_i^{t+1:T})$ can be used for additional smoothing of the temporal stochastic process, but is ignored during online processing.

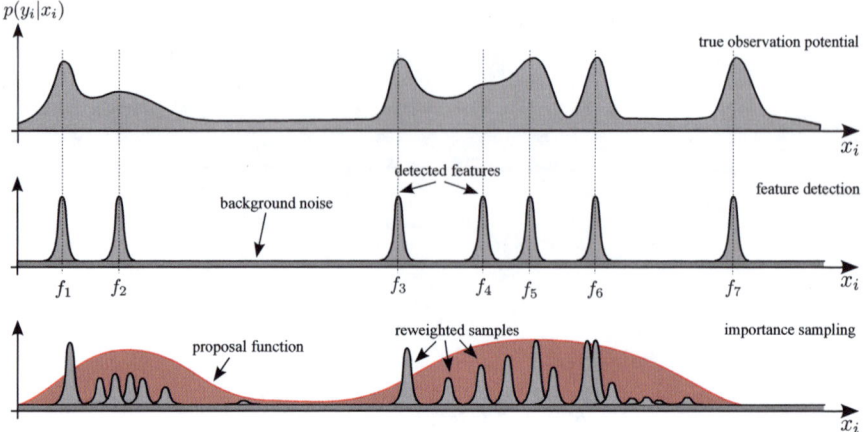

Fig. 3.6 Example of an observation potential (modified from [101]): true observation potential (top), approximation using feature extraction (center), approximation using importance sampling (bottom).

A comparison between different common approaches and their information sources is given in Fig. 3.5.

3.5 Observations

We will now refer to Sec. 2.1.4 and specify how observations are fed into the hierarchy. An observable node y_i is a noisy observation that is connected to a hidden random variable x_i via the observation potential $\psi(x_i, y_i)$. From a Bayesian point of view the observation potential is the likelihood $p(y_i|x_i)$ that describes how probable it is to observe y_i conditioned on x_i. A special feature of our approach is that we will consider two different ways to feed observations into the hierarchy. The traditional way is to gather observed values directly by means of feature detectors. Since the detected features represent noisy observations the observation potential has to be used to model uncertainties. We will further describe this approach in Sec. 3.5.1. Another solution offers importance sampling that is especially appropriate if no special detectors are available. Importance sampling and its application in this monograph will be introduced in Sec. 3.5.2. Depending on the measurement, different nodes in the hierarchy are appropriate to be linked to an observable node. These different scenarios are discussed in Sec. 3.5.3.

3.5.1 Feature Extraction

The data of images and image sequences is connected to a large visual input space and makes a direct processing infeasible. Feature extraction aims at reducing this space to a smaller set of features. One important issue during this extraction is that the relevant information from the input data should be kept while irrelevant information is reduced. Here, the type of relevant information is strongly task dependent and should be redefined for each new application. Furthermore, feature extraction is in many applications the first step of the image processing pipeline. Due to that, all subsequent processing steps rely sensibly on the quality of the first step and the overall image processing will often only be as good as its feature extraction.

Before discussing the properties of the feature extraction we have to clarify which type of features is used in the hierarchies. In general, there are two types we have to distinguish: local and global features. A *local feature* is characterized by a local image point, edge or region which differs from its immediate neighborhood in terms of changing image properties like intensity, color, and texture. *Global features* are calculated based on the whole image content. They ignore local properties like foreground and background information and simply combine all information. A typical example are color histograms, where the color distribution of the whole image is regarded as one global feature. Global features are typically used in the field of image retrieval or to gather contextual information. However, as we are interested in the spatial and spatiotemporal arrangements of the features, locality is an important information in our approach. Thus, we will in the following use the term 'feature' and always refer to local features. Nevertheless, global features might be helpful by providing contextual information, which could be integrated into our framework as described in Sec. 3.4. In this monograph we will however concentrate on local features.

A feature is defined as a pair (2-tuple) $f = (\mu, \delta)$ with a position $\mu \in \mathbb{R}^d$ with $d = 2$ for the image space and $d = 3$ for the space time volume, and a descriptor δ that describes the local immediate neighborhood around the feature position. Ideal features are characterized by the following properties [235]:

- Repeatability: The same features should be reliably detected in different images, taken under different viewing conditions.
- Distinctiveness/informativeness: In order to distinguish and match different feature types, the descriptors associated to the detected features should show strong variations.
- Locality: The features should be local, so that the region underlying the feature is small (reduces the influence of occlusions).
- Quantity: The number of detected features should reflect the complexity and the information content of an image.
- Accuracy: The detected features should be accurately localized.
- Efficiency: The detection process should be computational efficient especially for time-critical applications.

One problem arises concerning distinctiveness and locality since the two properties are competing [235]. If the features are built more locally the information content of the underlying intensity pattern decreases and thus leads to a lower distinctiveness. However, as we will use features in our hierarchy as local image evidence, we are more interested in good locality performance and thus accept a worse distinctiveness of the descriptor.

Feature extraction involves two stages. The first step is *feature detection* which is the actual information reduction step computing abstractions of the image information. The aim is to find those regions in an image or image sequence that show some local changes of the image properties like intensity, color, and texture. In general, feature detection algorithms are checking every image point sequentially and decide locally if there is a feature at that point or not. The outcome of the feature detection is a set of feature positions $\mu_1, ..., \mu_{N_f}$ representing a subset of the image domain. During the second step, *feature description*, the local neighborhoods around the feature positions are characterized by means of the descriptors $\delta_1, ..., \delta_{N_f}$. As already mentioned, the feature extraction is very task dependent. We will thus use different kinds of feature extraction techniques in the following chapters.

Feature extraction represents the input data as a collection of features $\mathcal{F} = \{f_1, ..., f_{N_f}\}$. An observable node i is associated with a feature appearance model containing a descriptor δ_i^c and has the local observation potential

$$\psi_i(x_i, y_i) = w_b \mathcal{N}(x_i; 0, \Lambda_b) + \sum_{n=1}^{N_f} w_n \mathcal{N}(x_i; \mu_n, \Lambda_f) \qquad (3.12)$$

with

$$w_n \propto \exp\left(-dist(\delta_n, \delta_i^c)^2 / 2\sigma^2\right) \qquad (3.13)$$

where $dist(\delta_n, \delta_i^c)$ is a distance metric between two descriptors. Since it is possible that a true feature has not been detected at all, we augment the Gaussian mixture model by a zero mean, high-variance $\Lambda_b \gg \Lambda_f$ Gaussian. This background Gaussian allows a good hypothesis to survive during the product calculation. The algorithms thus can handle outliers due to e.g. occlusions of the feature. The weight of the background Gaussian is calculated as

$$w_b = \frac{\lambda_b}{1 - \lambda_b} \sum_{n=1}^{N_a} w_n \qquad (3.14)$$

where λ_b determines the percentage of the background Gaussian. If $\lambda_b = 0.2$, it will represent 20% of the likelihood [226].

Let us for example consider the observation likelihood in Fig. 3.6 (top). A feature detector may gather a set of features as shown in Fig. 3.6 (center). The observation potential can be obtained by convolution of a Gaussian kernel with the discrete feature positions and adding a background level.

If every observable node would be associated with one individual feature appearance model, the computational effort would be unreasonably high. As mentioned, in our hierarchical framework we decide for locality and against distinctiveness of the descriptor. As a consequence the detected descriptors will in general not have many variations. Due to that, it is convenient to define a finite set of feature appearance models $\mathcal{C} = \{c_1, ..., c_{N_c}\}$, which are shared between the observable nodes. The finite set is often called a codebook or a vocabulary. A common way to determine an appropriate codebook is to apply a clustering algorithm to the descriptors of the training samples. Often, k-means clustering is used where the number of codewords has to be manually defined and an initial guess of the clusters has to be made. The clustering algorithm assigns each training sample to a cluster, where the cluster itself represents a descriptor, too, and is used as the codeword. Hence, the observable nodes of the hierarchy are assigned to one codeword $\delta_i^c = \delta^c$ of the vocabulary and the observation potential can be calculated as in eq. 3.12.

3.5.2 Importance Sampling

Another solution to feed observations into the hierarchy is to use *importance sampling* [16, 226, 7]. This technique applies when it is too complicated to sample from a probability distribution $p(x)$ directly. Supposing that a function $\pi(x)$ exists that is proportional to $p(x) \propto \pi(x)$ and that can be easily evaluated for any given value of x. Furthermore, we assume that the proposal $q(x)$ (also called *importance density*) is a function from which samples $x^{(n)} \sim q(x), n = 1, ..., N_s$ can be easily generated. The density $p(x)$ is then given by the weighted approximation

$$p(x) \approx \sum_{n=1}^{N_s} w^{(n)} \delta(x, x^{(n)}) \tag{3.15}$$

where

$$w^{(n)} \propto \frac{\pi(x^{(n)})}{q(x^{(n)})} \tag{3.16}$$

is the normalized *importance weight* of the nth particle. In practice, the chosen importance density critically influences the effectiveness of importance sampling [226]. In order to avoid as far as possible the generation of samples which are unlikely and have a low weight, the proposal function $q(x)$ should at least roughly approximate the shape of the desired distribution $p(x)$. Unlikely samples can be seen as wasted as they provide a negligible contribution to $p(x)$. On the other hand, the estimated distribution can be extremely inaccurate in regions of high importance of the target sample space if the proposal function $q(x)$ assigns low probability to these regions.

See for example the proposal distribution in Fig. 3.6 (bottom). The proposal density may be a mixture of two Gaussians (red curve). While the overall structure of the proposal matches the true density, especially the difference between the target density and the proposal on the right side of the function results in samples with negligible weights and misses to sample from the right peak of the target. Generally one can say, that a robust proposal distributions should be heavy–tailed and more dispersed than the target density [226, 138]. In the simplest case one choses an uniform proposal distribution.

We will now show, how importance sampling can be incorporated into graphical models. One popular example is the particle filtering approach where *sequential importance sampling* is applied [45, 226, 7]. The particle filter models a simple temporal stochastic process using a hidden Markov model (HMM) of order one. The idea of the HMM is to describe each discrete time step t by a hidden node $x_t \in \mathcal{X}_t$ representing the state vector and by an associated observable node y_t that is connected to the hidden state via the measurement model $p(y_t|x_t)$. The dependencies among the hidden nodes are specified by the state transition density $p(x_t|x_{t-1})$ between two subsequent time steps. Let us consider T points in time with the hidden states $\boldsymbol{x} = \{x_t\}_{t=0}^{T-1}$ and the observation sequence $\boldsymbol{y} = \{y_t\}_{t=0}^{T-1}$. The joint distribution then factorizes as

$$p(x, y) = p(x_0)p(y_0|x_0) \prod_{t=1}^{T-1} p(x_t|x_{t-1})p(y_t|x_t) \qquad (3.17)$$

As we are in general interested in the posterior distribution $p(x_t|\boldsymbol{y})$ of the state variable at the current time step t we can apply BP to solve the inference problem. The forward message that the hidden state x_t passes to state x_{t+1} can be recursively computed

$$m_{t,t+1}(x_{t+1}) \propto \int_{\mathcal{X}_t} p(x_{t+1}|x_t)p(y_t|x_t)m_{t-1,t}(x_t)dx_t \qquad (3.18)$$

This recursive formulation of the messages has an important interpretation for many tracking approaches. It allows to view each time step as comprising two stages [226]. In the first *prediction* stage, a forward message is calculated which corresponds to the predictive distribution of the state variable x_t given all previous observations $\boldsymbol{y}_{t-1} = \{y_0, .., y_{t-1}\}$

$$m_{t-1,t}(x_t) \propto p(x_t|\boldsymbol{y}_{t-1}) \qquad (3.19)$$

In the second *update* stage, the posterior filtered density is updated by combining the forward message with the new measurement at time step t

$$p(x_t|\boldsymbol{y}_t) \propto m_{t-1,t}(x_t)p(y_t|x_t) \qquad (3.20)$$

These two stages are the motivation for many tracking approaches. One characteristic of the particle filtering approach is that, as we are considering a continuous sample space, the posterior density is represented by a set $\{(x_t^{(n)}, w_t^{(n)})\}_{n=1}^{N_t}$ of weighted samples [226]

$$p(x_t|\boldsymbol{y}_t) = \sum_{n=1}^{N_t} w_t^{(n)} \delta(x_t, x_t^{(n)}) \qquad (3.21)$$

We can now apply importance sampling in order to get a density estimate $p(x_{t+1}|\boldsymbol{y}_{t+1})$ of the next time step. During prediction samples are generated from the proposal density $q(x_{t+1}|x_t, y_{t+1}) = p(x_{t+1}|x_t, y_{t+1})$

$$x_{t+1}^{(n)} \sim q(x_{t+1}|x_t^{(n)}, y_{t+1}) \qquad n = 1, ..., N_t \qquad (3.22)$$

The update stage can be seen as the evaluation of the likelihoods $p(y_{t+1}|x_{t+1}^{(n)})$ and as the calculation of the normalized importance weights

$$w_{t+1}^{(n)} \propto \frac{p(y_{t+1}|x_{t+1}^{(n)})p(x_{t+1}^{(n)}|x_t^{(n)})}{q(x_{t+1}^{(n)}|x_t^{(n)}, y_{t+1})} \qquad \sum_{n=1}^{N} w_{t+1}^{(n)} = 1 \qquad (3.23)$$

The proposal $q(x_{t+1}|x_t, y_{t+1})$ incorporates the subsequent observation y_{t+1} and provides an optimal proposal distribution which minimizes the variance of the importance weight conditional upon $x_t^{(n)}$ and y_{t+1} [45, 7]. However, in many tracking applications it is convenient to use the state transition density as the proposal $q(x_{t+1}|x_t, y_{t+1}) = p(x_{t+1}|x_t)$. In this case, just the dynamic model has to be simulated in order to predict the new state vector. Furthermore, the calculation of the importance weights equals the evaluation of the observations' likelihood $w_{t+1}^{(n)} \propto p(y_{t+1}|x_{t+1}^{(n)})$. Since these two steps are often simple to implement, the particle filter approach has reached a wide popularity in a wide range of applications.

We capture the idea of importance sampling and use it in our hierarchy. As mentioned before, one challenge is to make the high-level nodes observable. While it is difficult to sample from the likelihood $p(y_i|x_i)$ for high-level nodes directly, evaluation of the hypotheses is often tractable. We can thus build a sampling approach similar to the particle filter that first predicts the distribution of a hidden node based on the observation and the neighboring nodes (prediction) and then evaluates the samples using the observation potential (update). The proposal $q(x_i|\boldsymbol{x}_{\Upsilon(i)}, y_i)$ generates appropriate samples conditioned on the neighboring nodes $\boldsymbol{x}_{\Upsilon(i)}$ and the observation y_i. We choose the proposal $q(x_i|\boldsymbol{x}_{\Upsilon(i)}, y_i) = p(x_i|\boldsymbol{x}_{\Upsilon(i)})$ to factorize such that

$$q(x_i|\boldsymbol{x}_{\Upsilon(i)}, y_i) = \prod_{k \in \Upsilon(i)} p(x_i|x_k) \qquad (3.24)$$

This allows us to regard the proposal as the product of the incoming messages of node x_i. We can sample N new particles $\tilde{x}_k^{(n)}$ from the partial belief estimate

$$\tilde{x}_k^{(n)} \sim b_{i \backslash j}(x_k) \qquad (3.25)$$

and propagate it to the neighboring node x_i by sampling from $p(x_i|x_k)$ which represents the spatial or spatiotemporal relation between the nodes

$$x_i^{(n)} \sim p(x_i|\tilde{x}_k^{(n)}) \qquad (3.26)$$

Subsequently, the product of the messages has to be calculated (3.24) using one of the methods described in Sec. 2.3.2 or Sec. 3.8. In the update stage, the final weights can be evaluated

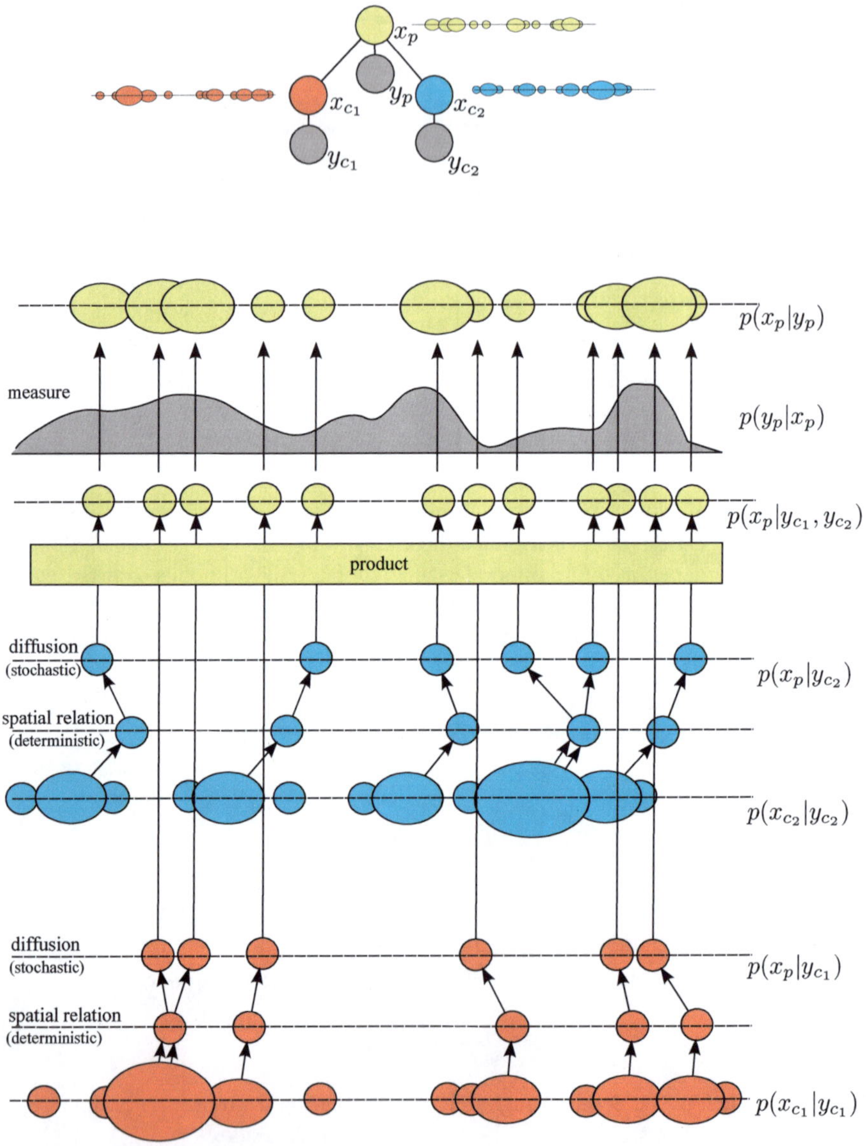

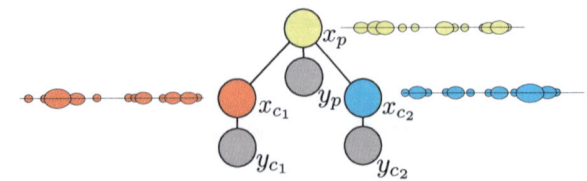

Fig. 3.7 Importance sampling for high-level nodes (modified from [101]). This figure illustrates how samples from the posterior marginal distribution $p(x_p|y_p)$ are generated by means of importance sampling. The low-level nodes x_{c_1}, x_{c_2} are propagating their estimates of the parent state to the parent. The product of the estimates is calculated. And finally, the samples of the product density are evaluated using the likelihood $p(y_p|x_p)$.

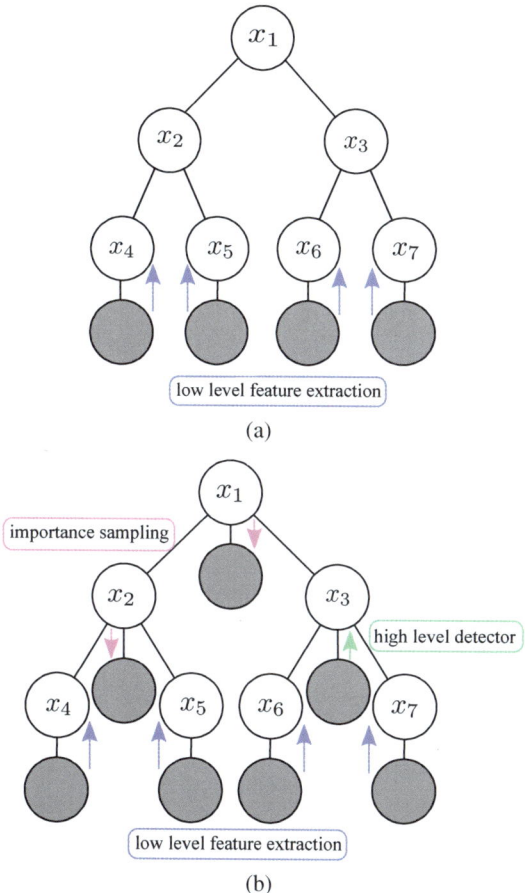

Fig. 3.8 Observable nodes (shaded) and hidden (open) nodes in a hierarchical graphical model. a) Just the low-level nodes are observable. b) All nodes are connected to an observable node.

$$w_i^{(n)} \propto p(y_i|x_i^{(n)}) \qquad \sum_{n=1}^{N} w_i^{(n)} = 1 \qquad (3.27)$$

A further simplification can be made during the bottom-up message propagation. In this stage, the message from the parent node is not available. Thus one can calculate an approximation of the proposal by using the messages from the children $q(x_i|\boldsymbol{x}_{\Upsilon(i)}, y_i) = p(x_i|\boldsymbol{x}_{\Xi(i)})$. An illustration of this importance sampling in a hierarchy can be seen in Fig. 3.7.

3.5.3   Observable Nodes in a Hierarchy

We will now briefly describe nodes of the hierarchy where measurements can be incorporated. The hierarchy is characterized by simple features representing the low-level nodes and by complex features representing the high-level nodes. In an early stage of the image processing and interpretation high-level information is generally not available. Hence, most common are hierarchies where just the low-level nodes are connected to observable nodes (see Fig. 3.8(a)) respresenting the early features and all other high-level nodes are hidden. The early features are detected as described in Sec. 3.5.1 and are used to start the bottom-up message passing. This is the approach, we will use in Chaps. 5, 6, and 7 of this monograph.

Another solution is to make all nodes observable, even the high-level ones (see Fig. 3.8(b)). On the one hand one could design special detectors that react on high-level stimuli like e.g. a face and feed this features into the hierarchy. However, this would require to train special detectors and to apply them to the input data causing an additional computational burden. On the other hand, the high-level feature information could also be gathered by other sensor modalities. In a vision-based surveillance system, the human body could be represented as described in Chap. 6 and could be detected based on simple early image features. Additionally, e.g. a presence sensor could be used to detect the person and the corresponding high-level information could be fed into the hierarchy. Although this raises interesting sensor fusion aspects, we do not further investigate this point. Importance sampling is another alternative to make the high-level nodes observable; it is especially suitable for hierarchies since evaluation functions can often be easily designed for high-level nodes (as described in Sec. 3.5.2). We will use this kind of observation in Chap. 8 and show that it significantly improves the accuracy of localization.

3.6   Spatial and Temporal Dependencies

The edge between two hidden nodes in the hierarchy defines a pairwise potential function. The random variable associated to each node has in general dimensions representing positional and orientational information, but other dimensions representing object properties like object width, height, or curvature are also possible. Let us first consider the positional part of a feature, which represents its pose in an image or image sequence. The potential function can therefore be interpreted as the spatial or spatiotemporal dependency between the nodes. One necessary step, before defining the potential function, is to assign a reference point or a reference coordinate system to each node. The potential function is then defined between these coordinate systems. Let x_i and x_j be two adjacent nodes, then their dependency is specified by the potential function $\psi_{ij}(x_i, x_j)$. In this monograph, the potential function depends only on the difference between neighboring variables $\psi_{ij}(x_i, x_j) = \tilde{\psi}_{ij}(x_i - x_j)$, hence, the marginal influence is constant and may be ignored (see Sec. 2.3.2). The potential functions are modeled as finite Gaussian mixtures. Since we are using the potential function in the nonparametric belief propagation framework, they are used

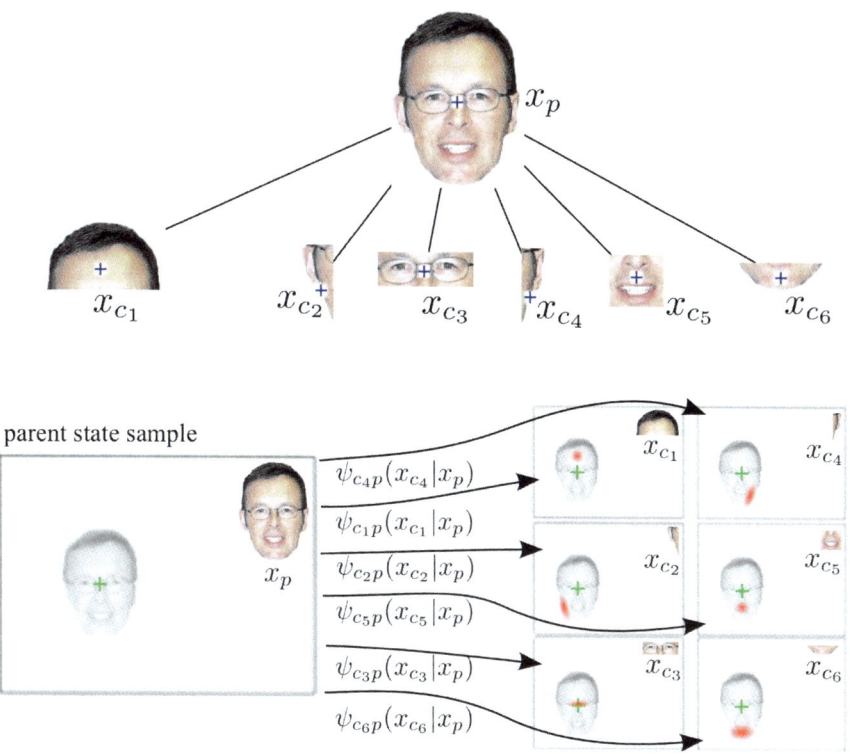

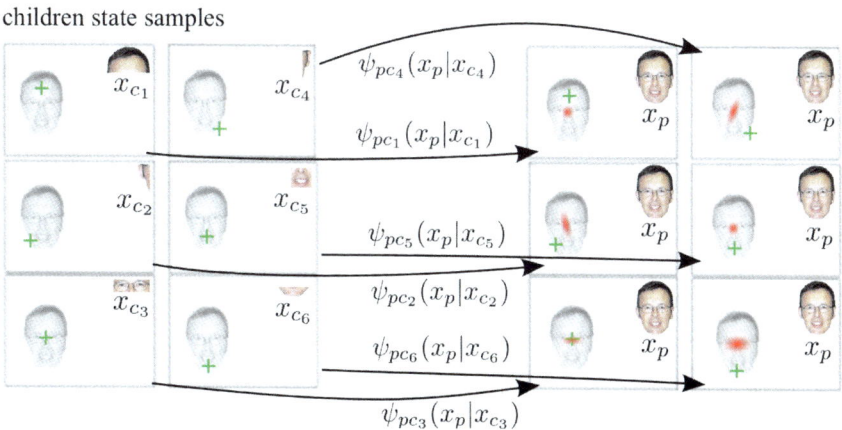

Fig. 3.9 Spatial relations between parent and children nodes. Hierarchy representing a face and its parts. The blue crosses are the reference coordinate systems (top), potential functions during top-down message passing (center), potential functions during bottom-up message passing (bottom) (face taken from Caltech-256 [83]).

as described in eq. 2.32 to draw samples $\tilde{x}_{ji}^{(n)}$ given the auxiliary particles $\tilde{x}_j^{(n)}$. We represent the potential as a mixture of Gaussian conditional distributions

$$\psi_{ij}(x_i, x_j = \tilde{x}_j^{(n)}) = \sum_{n=1}^{N_{ij}} w_n \mathcal{N}(x_i; \lambda_{jin}(\tilde{x}_j^{(n)}), \Lambda_{jin}) \qquad \sum_{n=1}^{N_{ij}} w_n = 1 \quad (3.28)$$

where $\lambda_{jin}(x_j)$ defines the spatial or spatiotemporal expected position of x_i based on x_j [204]. Let us for example consider a variable $x_j = (u_j, v_j, \theta_j) \in \mathbb{R}^3$ where (u_j, v_j) is the position in the image and θ_j the orientation angle. The variable x_j may be connected to a neighbor $x_i = (u_i, v_i, \theta_i) \in \mathbb{R}^3$. The spatial relation could in the simplest case be represented by relative position vectors $\boldsymbol{r}_{jin} = (r_{jin}^u, r_{jin}^v, 0) \in \mathbb{R}^3$, so that

$$\lambda_{jin}(x_j) = \begin{pmatrix} u_j \\ v_j \\ \theta_j \end{pmatrix} + \begin{pmatrix} \cos(\theta_j) & -\sin(\theta_j) & 0 \\ \sin(\theta_j) & \cos(\theta_j) & 0 \\ 0 & 0 & 1 \end{pmatrix} \boldsymbol{r}_{jin} \qquad (3.29)$$

The relative position vector $\boldsymbol{r}_{jin}$ is thus defined in the local coordinate system of x_j. As mentioned before, some dimensions may represent object properties. The formulation of the potential functions becomes especially challenging if there are statistical dependencies between the dimensions, so that the covariance matrix Λ_{jin} is not diagonal. Furthermore, if the dimensions of the parent are not equal to the dimensions of the child node, more complex potential functions have to be designed.

Since the potential function is used during message passing for each auxiliary particle of each node, the computational effort spent on sampling from eq. 3.28 significantly affects the performance. Therefore it is preferable to use simple potential functions which can be easily evaluated and from which samples can be easily drawn. This is one reason why we will suggest in Sec. 3.7.1 to use random variables which just model the 2d image position. If we additionally simplify the mixture of Gaussian to a simple Gaussian model with one relative position vector $\boldsymbol{r}_{ji}$, the potential function becomes

$$\psi_{ij}(x_i, x_j = \tilde{x}_j^{(n)}) = \mathcal{N}(x_i; (\tilde{u}_j^{(n)}, \tilde{v}_j^{(n)}) + \boldsymbol{r}_{ji}, \Lambda_i) \qquad (3.30)$$

An example can be seen in Fig. 3.9. Here, the 2d relations between a face and its parts are shown. As can be seen, the potential functions are defined by conditional Gaussian models depending on the direction of the message passing, top-down message passing (center) and bottom-up message passing (bottom).

In order to further decrease the computational effort we restrict the covariance matrix to be diagonal, so that the potential can be factorized as

$$\psi_{ij}(x_i, x_j = \tilde{x}_j^{(n)}) = \mathcal{N}(u_i; \tilde{u}_j^{(n)} + r_{ji}^u, \sigma_i^u)\mathcal{N}(v_i; \tilde{v}_j^{(n)} + r_{ji}^v, \sigma_i^v) \qquad \Lambda_i = \begin{pmatrix} \sigma_i^u & 0 \\ 0 & \sigma_i^v \end{pmatrix} \tag{3.31}$$

Interestingly, we can use the same concept for the representation of activities. We just have to regard random variables $x_j = (u_j, v_j, t_j) \in \mathbb{R}^3$ where (u_j, v_j) is the image position, and t_j is the time step when the activity occurred. Then, the potential function can be expressed by means of a 3d relative position vector $\boldsymbol{r}_{ji}$

$$\psi_{ij}(x_i, x_j = \tilde{x}_j^{(n)}) = \mathcal{N}(x_i; (\tilde{u}_j^{(n)}, \tilde{v}_j^{(n)}, \tilde{t}_j^{(n)}) + \boldsymbol{r}_{ji}, \Lambda_i) \tag{3.32}$$

We will further describe this idea in Chap. 7.

3.7 Compositional Hierarchical Models

In this section, we will motivate and introduce the main concept of our new visual hierarchical model, that we will apply in Chap. 5 to achieve object recognition and in Chap. 6 to detect the human body. Modified versions will also be applied to activity recognition in Chap. 7 and scene understanding for intelligent vehicles in Chap. 8.

Our model is represented by a set of hierarchies (also called a forest of trees) $\mathcal{G} = \{\mathcal{G}_1, ..., \mathcal{G}_N\}$, which are undirected tree-structured graphs $\mathcal{G} = (\mathcal{V}, \mathcal{E})$ as described in Sec. 3.1 and a set of additional edges $\mathcal{L}$, which represents links between the different graphs. These links are used to share information among different graphs according to the similarity edges introduced in Sec. 3.4. Each tree represents a compositional hierarchy.

Definition 1 (Compositional Hierarchy). *A compositional hierarchy is a rooted tree, which defines the decomposition of an object, represented by the root, into smaller parts, represented by the children.*

The nodes of the compositional hierarchy are associated to features, compound features, parts or objects in the image space $\Omega \subseteq \mathbb{R}^2$. Each of these elements is represented by a set of features $\mathcal{F} = \{f_1, ..., f_{N_f}\}$, where the features are described by their local appearance descriptors δ_i and positions μ_i. A reference point is used to precisely describe the position of a feature set. In principle, this reference point can be arbitrarily chosen. However, in this monograph, we will use the center $r = 1/N_f \sum_{i=1}^{N_f} \mu_i$ as the reference point. The idea of a compositional hierarchy is that the feature sets are iteratively decomposed into smaller sets, i.e. the feature set of a parent node $\mathcal{F}_p$ is decomposed into feature sets of the children $\mathcal{F}_c$ with $\mathcal{F}_p = \mathcal{F}_{c_1} \cup \mathcal{F}_{c_2} \cup ... \cup \mathcal{F}_{c_n}$.

Definition 2 (Structure of a Compositional Hierarchy). *The structure of a compositional hierarchy is unambiguously defined by the decomposition of each feature set.*

During decomposition, the nodes are assigned to a hierarchy level according to their size.

Definition 3 (Level of a Node). *The level of a node is directly determined by the number of features of its associated feature set $\ell = \ell(N_f)$.*

We will in the following use

$$\ell(N_f) = \text{round}(\log_{\sqrt{2}} N_f) \qquad (3.33)$$

Thus, the nodes at higher levels represent more complex parts or objects, while the low-level nodes represent simple features and compound features. One advantage of the direct coupling between the feature size and hierarchy level is that the nodes of one level have the same complexity and can be compared more easily. Please note that as a consequence the node's height in the tree does not have to correspond to the node's level in the hierarchy. Thus, not every level under the root node has to be used.

3.7.1 Sets of Hierarchies

A compositional hierarchy defines the decomposition of an object into smaller parts. The spatial relations between the objects parts (encoded in the pairwise potential functions) are thereby allowing a certain degree of spatial variations. However, as soon as the variations become too large or other objects have to be represented, new separate compositional hierarchies have to be defined. This results in a set of hierarchies $\mathcal{G} = \{\mathcal{G}_1, ..., \mathcal{G}_N\}$, where each hierarchy represents one specific object at one specific view, articulation, scale, and orientation. At first glance, this representation may seem unattractive: even for a small number of values for each dimension, the total number of instances is large due to the high dimensionality (view, articulation, scale, orientation). However, the modeling of translation, scale and rotation as separate hierarchies has several advantages. It allows us to efficiently reduce redundant calculations, since the reusability of the parts between the different configurations is very high. Furthermore, the complexity of structure learning is reduced since all instances for one specific view and articulation share the same hierarchical decomposition. Just the relative scales and angles have to be adapted. We therefore group all elements to one view-dependent representation of an object and share the structure among them. Furthermore, as we will see in Sec. 3.7.4, this representation also allows us to construct an efficient coarse-to-fine extension.

3.7.1.1   View-Dependent Representation of an Object

Objects are represented by view-dependent instances. The overall appearance of these view-dependent instances stays constant during translation in x, y direction, scaling and 2d rotation. This is an important property since it allow us to use and to share the same hierarchical decomposition for all of these configurations. During scaling we have just to scale and during rotation to rotate the potential functions.

We map the continuous scale s and rotation θ to their quantized counterparts $\hat{s}$ and $\hat{\theta}$ with a finite number N_s and N_θ of values. The scale and rotation space can then be represented by a set of instances $x^{vd,o} = \{x^{vd,o}_{\hat{s},\hat{\theta}}\}_{\hat{s}=1,...,N_s,\hat{\theta}=1,...,N_\theta}$. Each hierarchy has one root that is associated to one specific scale $\hat{s}$ and rotation

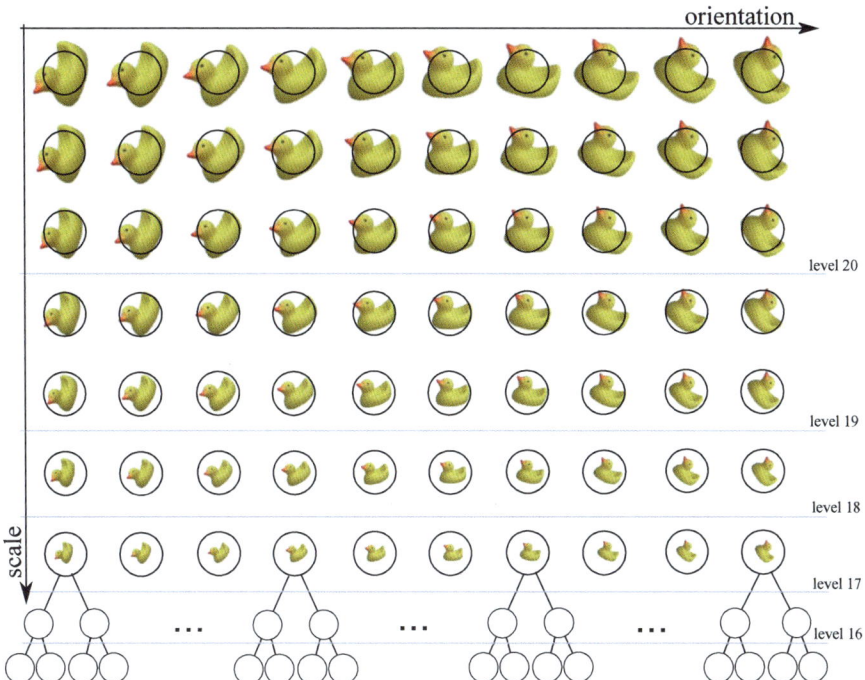

Fig. 3.10 View-dependent representation of an object as a set of hierarchies. Each scale and rotation is represented by a separate compositional hierarchy (object taken from COIL-100 [165]).

$\hat{\theta}$ instance, and the corresponding feature set $\mathcal{F}_{\hat{s},\hat{\theta}} = \{f_{\hat{s},\hat{\theta},1}, ..., f_{\hat{s},\hat{\theta},N_{\hat{s},\hat{\theta}}}\}$. Since $\mathcal{F}_{\hat{s},\hat{\theta}}$ is independent w.r.t. translation in x,y direction, we will represent these two dimensions by one node and encode it in the random variable $x_i \in \mathbb{R}^2$ associated to node i. Consequently, all other nodes will also be associated with a two-dimensional random variable leading to simple spatial pairwise potential functions as discussed in Sec. 3.6. The quantized scales and rotations are separately encoded as nodes in the hierarchy. Thus, each node of the hierarchy is a random variable $x_{\hat{s},\hat{\theta}}^{vd,o} \in \mathbb{R}^2$ associated to a specific scale $\hat{s}$ and rotation $\hat{\theta}$. The nodes for a given scale $\hat{s}$ are on the same hierarchy level since the number of features during rotation stays constant. This is in general not the case if the scale is changed. Whenever the number of features exceed or fall below the level thresholds $\tau_\ell = 2^{l/2}$ with $l = 1, ...N_l$ the level of the associated node changes.

Definition 4 (Viewpoint Dependent Representation). *The viewpoint dependent representation of object o is given by the set $x^{vd,o} = \{x_{s,\theta}^{vd,o}\}_{s \in S, \theta \in \Theta}$ of scale and orientation variant models.*

As depicted in Fig. 3.10, each root node represents an object with a specific scale and orientation. For clarity reasons just the hierarchies of the instances at the bottom are shown. Please note that we construct hierarchies for all scales and rotations of a view-dependent instance, and we, hence, get automatically a dense scale and rotation representation for all parts of the instance.

This representation is also biologically inspired. One widely accepted fact about the ventral stream in the visual cortex is that visual processing is hierarchical [210, 211]. The hierarchical structure aims at building invariance to position, scale and rotation first. Upon this translation, scale and rotation invariant representation, viewpoint and other transformations are modeled. That is what inspired our idea of modeling object configurations with compositional hierarchies.

3.7.1.2 View-Independent Representation of an Object

Because it is difficult to analytically describe the spatial and geometrical relationships between parts, and furthermore 3d models are often not available, we use the view-dependent object instances to get a view-independent representation of an object. This representation fits well into the forest structure discussed previously. We built for every viewpoint a view-dependent representation according to the quantized viewpoints $\hat{v} = \{\hat{v}_1, ..., \hat{v}_{N_v}\}$ as described in Sec. 3.7.1.1:

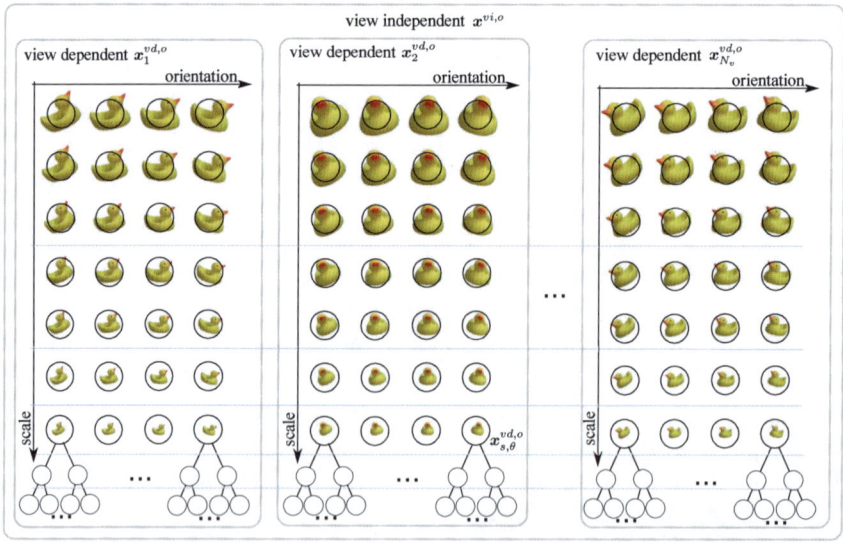

Fig. 3.11 View-independent representation of an object as a set of view-dependent models (object taken from COIL-100 [165]).

Definition 5 (View-Independent Representation). *The view-independent represen-tation of an object o is given by the set $x^{vi,o} = \{x_{\hat{v}_n}^{vd,o}\}_{n \in N_v}$ of viewpoint dependent models.*

As illustrated in Fig. 3.11, the object representation maps the quantized view-points $\hat{v}$ to groups of root nodes. An object representation $x^{vi,o}$ is represented by a set of view-dependent models $x^{vd,o}$, these in turn, are represented by a set of scale and rotation dependent root nodes. Note that the root nodes for one specific scale and rotation, but arbitrary view-points, need not necessarily be on the same hierarchy level. This is especially the case for objects, where a textured side has an increased number of features, so that according to eq. 3.33 the level will be higher.

It is also worth noting, that the different viewpoints, scales or rotations do not have to be uniformly distributed. In most applications a prior distribution for the configurations is available. Such distributions could e.g. define constraints in the configuration space due to interaction of the object with the environment with which it comes into contact (for example a book lying on the table). The prior distribution could also be defined by a density distribution, which assigns high density values to important regions of the configuration space and low values elsewhere. Although this variable sampling from the configuration space sounds promising, it will not be discussed in any further detail in this monograph.

There is also biological evidence for this view-dependent approach as shown by Tarr et al. [230]. In an experimental setup, they found that recognition of sin-gle primitives is progressively more difficult as the difference between studied and tested viewpoints increased. From that findings they concluded that 3d object recog-nition in the human visual system is based on view-based representations and recog-nition processes.

3.7.1.3 Representation of Multiple Objects and Articulated Objects

The view-independent representation can be extended to articulated objects or to multiple object classes. In either case, the principle stays the same. For each object and articulation a view-independent representation has to be constructed. As before, the continuous dimensions of articulations a have to be mapped to their quantized counterparts $\hat{a}$ with a finite numbers N_a of values.

Definition 6 (Multi-Object Representation). *The representation of multiple objects $o = \{o_1, ..., o_{N_o}\}$ at different articulations $\hat{a} = \{\hat{a}_1, ..., \hat{a}_{N_a}\}$ is given by the set $x^{vi,o} = \{x^{vd,o_i,\hat{a}_j}\}_{i \in N_o, j \in N_a}$ of viewpoint dependent models.*

3.7.2 Similarity between Nodes

In this section, we will introduce an additional set of edges, which is used to connect the trees with each other. Two nodes of the same hierarchy level, which may belong to the same object or not, are connected according to their similarity. Similarity is defined by means of the associated feature set.

Definition 7 (Similarity Criterion). *Let $\mathcal{F}_i$ denote the feature set associated to node i and $\mathcal{F}_j$ that of node j. An edge e_{ij} is added to the set of edges $\mathcal{L}$, if the distance between the feature sets is less than a threshold $dist(\mathcal{F}_i, \mathcal{F}_j) < \tau_s$.*

The potential function associated to edge e_{ij} is then defined by a mixture of Gaussian conditional distributions

$$\psi_{ij}(x_i, x_j) = \alpha_s \mathcal{N}(x_i; x_j, \Lambda_s) + (1 - \alpha_s)\mathcal{N}(x_i; 0, \Lambda_b) \tag{3.34}$$

with

$$\alpha_s = \exp\left(-\xi_s \left(dist\left(\mathcal{F}_i, \mathcal{F}_j\right)\right)^2\right) \tag{3.35}$$

The parameter α_s controls the influence of the nodes on each other. If the two feature sets are equal $dist(\mathcal{F}_i, \mathcal{F}_j) = 0$, $\psi_{ij}(x_i, x_j)$ corresponds to a Gaussian conditional distribution with mean x_j and a diagonal covariance matrix Λ_s with small variances. On the other hand, if $\alpha_s \approx 0$, $\psi_{ij}(x_i, x_j)$ will correspond to a background Gaussian conditional distribution with a high variance Λ_b which leads to no influence.

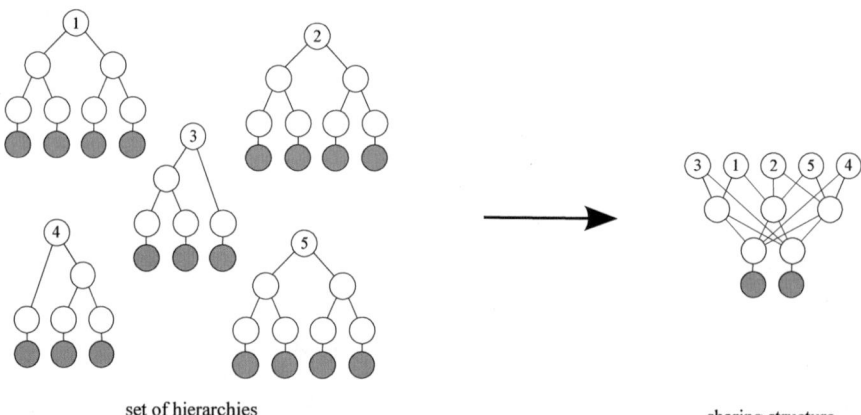

set of hierarchies   sharing structure

Fig. 3.12 Set of hierarchies and sharing structure for five configurations. The hierarchies are separately modeled (left). Efficient sharing on all hierarchy levels (right). Please note, the number of root nodes stays the same.

3.7.3 Sharing

Up to now, the object is represented by a forest of trees $\mathcal{G} = \{\mathcal{G}_1, ..., \mathcal{G}_N\}$, where each tree represents one specific scale, rotation, view or articulation of one specific object (see Fig. 3.12 (left)), and a set of links between the trees $\mathcal{L}$. As mentioned before, this leads to an exponential growth of the number of instances and makes inference as well as learning intractable. Fortunately, many nodes are equal or at least similar. This similarity can be used to share information between the nodes and reduce the complexity of inference. In Sec. 3.7.2 we introduced a threshold τ_s to decide if a similarity link between two nodes has to be established. We use a second threshold τ_e with $0 \leq \tau_e \leq \tau_s$ to decide if two nodes are equal or not.

Definition 8 (Sharing Criterion). *Two nodes are equal and thus can be shared, if the distance between the feature sets is less than a threshold:*

$$dist(\mathcal{F}_i, \mathcal{F}_j) < \tau_e \qquad (3.36)$$

Often groups $\mathcal{S}$ of two or more similar nodes exist, which all fulfill the sharing criterion among each other.

Definition 9 (Sharing Group). *A group $\mathcal{S}$ of nodes is called a sharing group if all nodes fulfill the sharing criterion among each other.*

In order to reduce the complexity of the hierarchical structure we represent a sharing group by one representative node r.

Definition 10 (Representative of a Sharing Group). *The representative of a sharing group $\mathcal{S}$ is defined as:*

$$r = \arg\min_{i \in \mathcal{S}} \sum_{j \in \mathcal{S}} dist(\mathcal{F}_i, \mathcal{F}_j) \qquad (3.37)$$

All elements of a sharing group share the structure as well as the local evidence of the representative.

When we consider the different information sources as described in Sec. 3.4, then similar nodes can use the potential $\psi_{ij}(x_i, x_j)$ to share information among them and improve the robustness of the classification. However, this does not reduce the number of nodes. A better solution is therefore to merge equal nodes during inference. Similar nodes have a similar appearance and also similar local evidence. Thus, the local evidence that reaches the nodes over the children $p(x_i|\boldsymbol{y}_\Xi)$ is approximately equal. On the other hand, the information from the parents is in general different since the geometrical/spatial context is different. A wheel of a car shares for example its circle shape with an eye of a face. While the local evidence of these two object classes is similar, the context is different. The wheel appears in the context of a car with a specific spatial distribution, the eye appears in the context of a face. It follows, although equal nodes share local evidence and their compositional hierarchical structure, they still get different contextual information from their parents.

During inference, we can therefore schedule the message passing as described in Sec. 2.3.3, and divide the computation into two stages. In the upward sweep, the messages are propagated from leaves to the root. In this stage the local evidence that reaches the nodes over the children nodes $p(x_i|\boldsymbol{y}_\Xi)$ is calculated. We can therefore share information and avoid redundant calculations. Instead of calculating the partial belief estimate separately for each of the equal nodes, the belief is just calculated

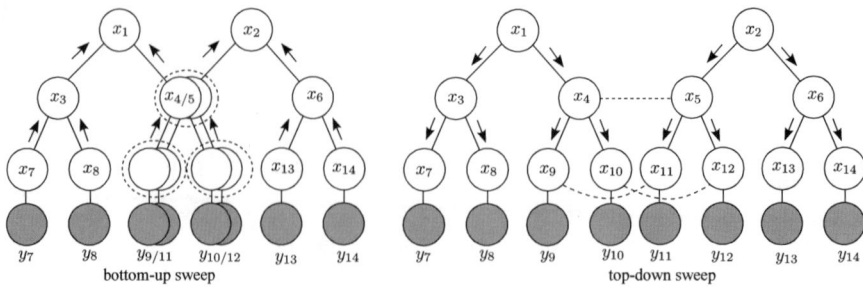

Fig. 3.13 Linked trees: The graphs demonstrate the usage of the similarity links during message passing. The root nodes x_1 and x_2 represent two configurations, which share the primitive node $x_{4/5}$. During the upward sweep messages are passed from bottom to top. Here, the link between node x_4 and x_5 is used to share the belief among each other (left). However, during the downward passing the links are not used any more since the nodes are parts of different root nodes (right).

once and sent to all equal nodes and their parent nodes. This sharing is illustrated in Fig. 3.13 (left), where the nodes $x_{4/5}$ are shared. In the second stage, the downward sweep, the messages are propagated from the root back to the leaves. The links are not used any more (see Fig. 3.13 (right)). Please note, although the linked graphs in Fig. 3.13 (right) look like a polytree, every graph is still tree-structured. Thus, the upward and downward ordering of the updates requires each message to be computed only once. If we would not remove the links during the downward sweep, the graph could have loops and thus convergence could not be guaranteed [159].

Another advantage is that sharing facilitates the learning procedure since the hierarchical compositional structure of equal nodes has to be learned just once. Here, the right choice of the threshold τ_e is crucial for the performance as well as for the efficiency as illustrated in Fig. 3.12. If the threshold is too low no primitives are shared (left). The higher the threshold τ_e, the higher the number of shared nodes (right). However, higher thresholds lead simultaneously to inaccurate and distorted representations. In borderline cases, root nodes are merged preventing to distinguish between them resulting in worse discriminative performance.

The previously introduced forest contains separately modeled hierarchies, which are linked according to their similarity as discussed in Sec. 3.7.2. In an ideal probabilistic framework, the densities of all nodes in the hierarchy would be exact. In this case we can expected that similar nodes will have similar distributions without explicitly sharing information. However, in our framework the configuration space is quantized, which complicates the generalization, and in addition the distributions are nonparametrically represented by a finite set of particles. This finite set necessitates that information is shared between similar nodes in order to increase the generalization properties and to increase the robustness of the recognition process. Similarity edges are also discussed in [202, 64]. Here, similarity edges between views and between parts within layers and across layers are used for greater generalization. Since the compositional hierarchies in [64] are quite simple and have

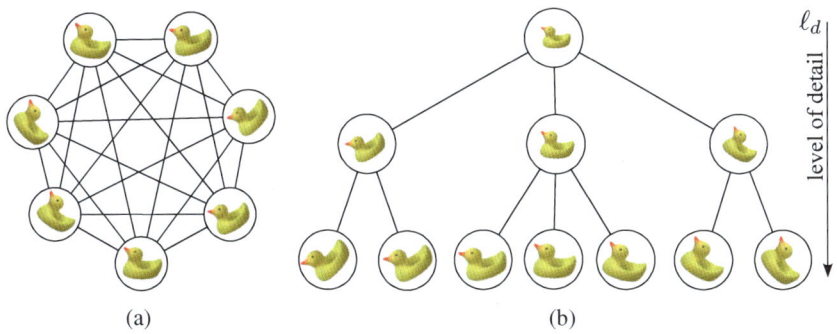

Fig. 3.14 Similarity between nodes: a) Fully connected. b) Similarity hierarchy.

just four levels, the direct establishment of similarities between hierarchical nodes might be tractable. However, in our hierarchy the number of similar objects and parts is very high due to the quantization of the configuration space. Defining similarities between nodes induces large numbers of edges resulting in fully connected subgraphs as depicted in Fig. 3.14(a). This leads to an inefficient representation of the hierarchy and also makes inference difficult since a large number of messages has to be calculated. Furthermore, it is challenging to implement an efficient message passing schedule. Since the similarity edges cause loops convergence of the message passing cannot be guaranteed.

3.7.4 Coarse-to-Fine Hierarchy

We propose to use a similarity hierarchy as depicted in Fig. 3.14(b) to share information between similar nodes. The idea is to group nodes according to their similarity and represent them by a parent node. The parent node guarantees that information between the children can be shared although they are not directly connected. It can thus be seen as a generalization of its children that is less specific and has a reduced level of detail.

Definition 11 (Similarity Hierarchy). *A similarity hierarchy connects a node x with level of detail ℓ_d and compositional level ℓ to similar nodes on a coarser $\ell_d - 1$ and a finer $\ell_d + 1$ level.*

At this point we have to emphasize the difference between similarity hierarchies and compositional hierarchies. While the compositional hierarchy combines parts or features to build more complex parts or objects, the similarity hierarchy models the similarity between parts. The potentials associated to the edges are those discussed in Sec. 3.7.2 and are not spatial relations as discussed in Sec. 3.6.

The parent nodes in the similarity hierarchy are generalizations of their children with a reduced level of detail. We combine this generalization with a scale space representation, where the level of detail is implicitly connected to the scale level.

The scale space is a multi resolution representation of an image [131, 132]. It can be thought of as a set of images which are all based on the same input image but with different levels of detail. Formally, the scale-space representation $S : \mathbb{R}^2 \times \mathbb{R} \to \mathbb{R}$ of the image $I : \mathbb{R}^2 \to \mathbb{R}$ is defined by a convolution

$$S(x;t) = \int_{\xi \in \mathbb{R}^2} I(x - \xi)g(\xi;t)d\xi \qquad (3.38)$$

where $g(x;t)$ denotes a Gaussian kernel with variance t^2 used for image blurring. The discrete scale space, which is sampled in space and scale, is represented by an image pyramid, in which the resolution is downsampled by $\sqrt{2}^{-1}$.

Definition 12 (Scale Space Representation). *The level of detail ℓ_d corresponds to the discrete scale $s_{\ell_d} = 2^{\ell_d/2}$ in a scale space.*

We establish the two kinds of hierarchies pictorially in horizontal and vertical directions (see Fig. 3.15).

Definition 13 (Compositional Hierarchies at Different Levels of Detail). *The nodes of one level of detail ℓ_d are forming a compositional hierarchy, the similarity hierarchies connect compositional hierarchies at different levels of detail.*

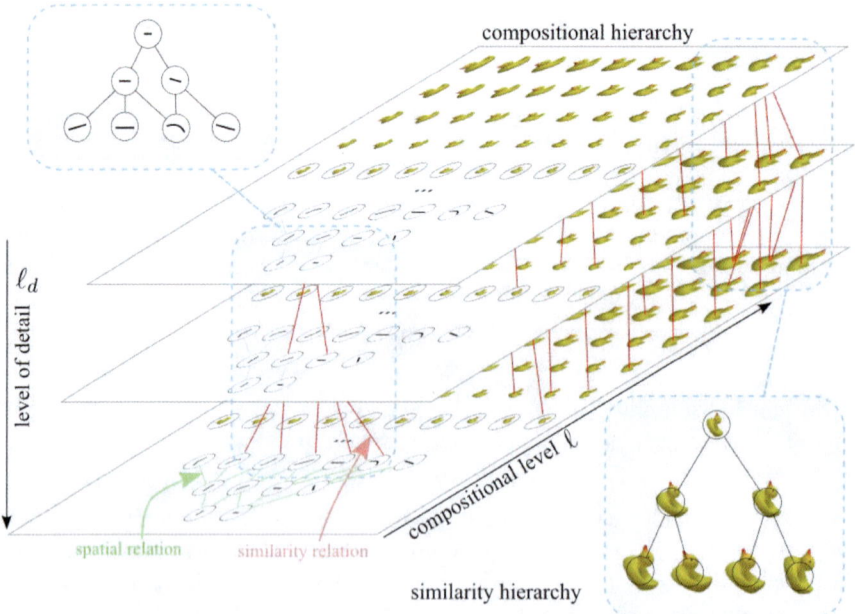

Fig. 3.15 Combined compositional and similarity hierarchy. In horizontal direction the compositional hierarchies are arranged, while in vertical direction the similarity hierarchies are constructed.

In Fig. 3.15, the compositional hierarchies are arranged in horizontal direction, while the similarity hierarchies are constructed in vertical direction. This structure gives us a representation of the compositional hierarchies at different levels of details. While the bottom level models the object at the original scale, higher levels can be used to represent the object at lower levels of detail.

Using the scale space representation facilities the construction of the similarity hierarchy. We can use the fact, that the compositional hierarchy already contains scaled versions of each node. For each node we have a node at a scale on the next lower level that simultaneously represents the object at a coarser level of detail. Since the scale space is downsampled by $\sqrt{2}^{-1}$ and the size of the feature set changed by the factor $\sqrt{2}^{-1}$ as well, we can use the compositional hierarchy, shift it by one level and use it as a hierarchy at a coarser level of detail.

3.7.4.1 Inference in a Coarse-to-Fine Hierarchy

There are two reasonable schedules for the message passing, which use the similarity edges to share information between nodes. The first one proceeds, as before, at the finest level of detail. The low-level features are fed into the hierarchy and messages are sent horizontally to the parent nodes. The vertical hierarchies can be seen as auxiliary hierarchies, which are used to share information. During message passing nodes at higher vertical levels can be inferred based on already detected children nodes. The parent node can then send information back to the nodes at the bottom.

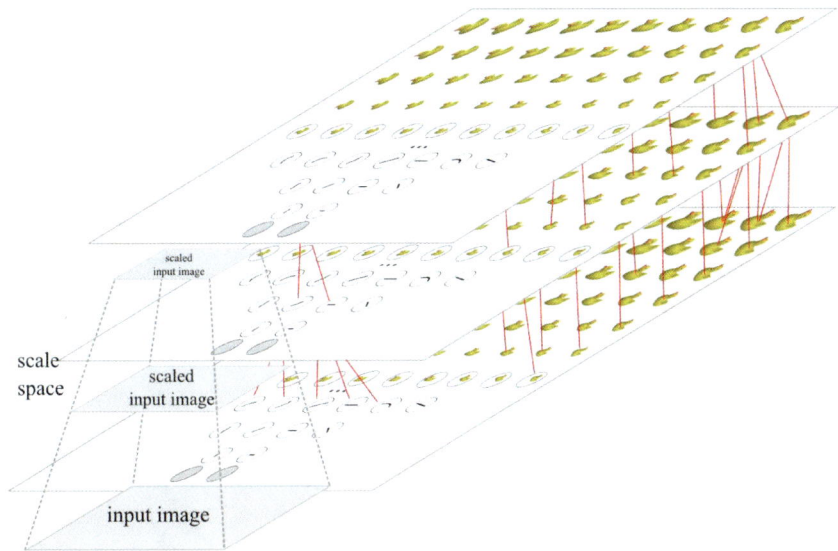

Fig. 3.16 A compositional and similarity hierarchy with a scale space representation

However, this schedule still requires to perform inference at the finest level of detail. This is time consuming since during bottom-up inference a lot of calculations and time is spent on evaluation of small structures that often just represent textures. Therefore, these small structures are detected before large objects. Unfortunately, in most applications the reverse order would be desirable. We therefore combine the structure with a coarse-to-fine approach: we perform horizontal inference at coarser levels and refine the solutions by searching in vertical direction. The number of nodes at the higher levels is reduced, so that less messages have to be calculated during inference. The coarse-to-fine approach is also attractive, since it can be directly combined with a scale space representation of the input image as depicted in Fig. 3.16. Since all levels of detail are built of the same compositional hierarchy (see the previous section), they all contain the same observable low-level features. Thus, we can apply the feature detector to different scales of the input image and directly feed observations into the model at coarser levels. The idea is (see Fig. 3.17):

1. to detect objects at a coarser level of detail (proceeding horizontally bottom-up)
2. to refine the detected solutions (proceeding vertically top-down)
3. and to evaluate the solutions (proceeding horizontally top-down)

The first step corresponds to a hypothesis generation step, where the compositional structure at a coarse level of detail is used to share hypotheses and efficiently index possible candidates. In the second step, the hypotheses are refined by proceeding

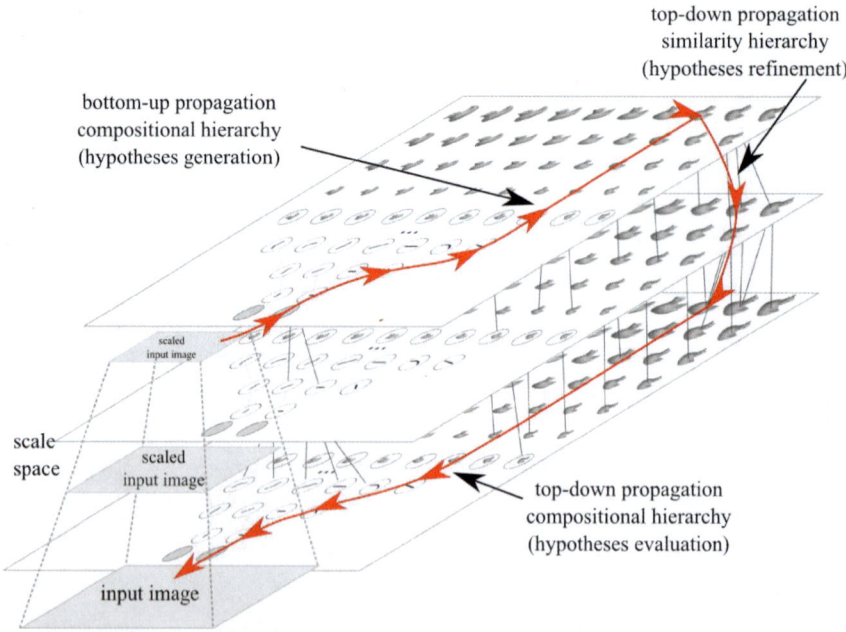

Fig. 3.17 Message passing in a compositional and similarity hierarchy

top-down in the similarity hierarchy. Here, each node sends a message to all of its similar nodes at the next finer resolution. The refined hypotheses are then evaluated by proceeding top-down in the compositional hierarchy. In order to reduce the complexity of the search and the number of hypotheses, it is reasonable to use just the best refined hypothesis according to the evaluation. This hypothesis is then again propagated to the next finer level, evaluated, and so on. The refinement and evaluation steps are continued until finally the finest level of detail (the original image size) is reached. In order to detect also small objects, the first step should be applied to several scaled input images. The idea of propagating single hypotheses or particles through the graph will be further discussed in Sec. 3.8.2.

3.8 Inference

In this section we will briefly summarize the inference steps discussed in the previous sections and point out some challenges like e.g. handling of occlusions. For clarity reasons we will consider horizontal hierarchies and ignore vertical dependencies, which are discussed in Sec. 3.7.4.1. The message passing starts at the bottom, where the low-level information gathered by the feature detectors is fed into the hierarchy. Due to the sharing structure discussed in Sec. 3.7.3 messages will typically be sent to several parent nodes. Once a node receives the messages from its neighbors it can estimate its belief (or partial belief if not all message are received) by calculating the product. As already mentioned in Sec. 2.3.2, the calculation of this product is very challenging since the resulting mixture will be comprised of Ld Gaussians, if each of the d mixtures has L Gaussians. Sudderth et al. [227] proposed to use the Gibbs sampler to efficiently sample from the product. However, we found that this sampler has difficulties to sample from multimodal distributions. The number of internal iterations κ has to be high in order to get accurate samples. Unfortunately, this results in high computational costs. We therefore propose in Sec. 3.8.1 two new efficient product calculation methods that show excellent performance in our experimental evaluation. Furthermore, we will in Sec. 3.8.2 combine the bottom-up message passing with a top-down evaluation step.

3.8.1 Efficient Products of Gaussian Mixtures

We will now propose two efficient product calculation methods that are especially suitable for real-time applications. The first method reduces the number of particles and performs an exact product calculation on the simplified particle sets. The second method approximates the product by an efficient nearest neighbor search.

3.8.1.1 Exact Products of Gaussian Mixtures Using Reduction Methods

The exact calculation of the product is still preferable since it regards all modes of a multimodal density. However, for large particle sets the exact computation is intractable due to the combinatorial complexity. Fortunately, a significant number of the particles contribute little to the density estimate and are often similar to other particles. One can therefore simplify the densities and reduce the number of particles. Exact product calculation can then be applied to the reduced particles sets. Jeon and Landgrebe [102] proposed to use a pre-clustering of the data and a simple branch and bound procedure to significantly reduce the numbers of data samples which would contribute little to the density estimate. Their technique is especially helpful in the multivariate case, and does not require an uniform sampling grid. Babich and Camps [8] proposed a similar clustering approach to find a set of reference vectors and weights which are used to approximate the Parzen-window classifier.

We use a simple but efficient density-based clustering algorithm [52] to find a reduced particle set. The advantage is that it can find arbitrarily shaped clusters and requires just two parameters: the maximal neighborhood distance ε and the minimum number $n_{\min}$ of points required to form a cluster. The approach starts with an arbitrarily chosen particle $x_j^{(n)}$ and retrieves all particles that are within the ε neighborhood. If the number of found neighbors is greater than $n_{\min}$, a cluster is initiated and the particle is assigned to this cluster. Otherwise, the particle is labeled as noise. This neighborhood check is recursively repeated for all neighbors. If a particle is found to be a ε neighbor, its neighborhood will also be assigned to the same cluster. This procedure is repeated for all unvisited particles. After the algorithm terminates, the clusters define the reduced particle set. We represent the cluster $\mathcal{C}_i$ by a new weighted particle $(\tilde{x}_j^{(i)}, \tilde{w}_j^{(i)})$ that corresponds to the cluster's center:

$$\tilde{x}_j^{(i)} = \frac{1}{\tilde{w}_j^{(i)}} \sum_{n \in \mathcal{C}_i} w_j^{(n)} x_j^{(n)} \qquad\qquad \tilde{w}_j^{(i)} = \sum_{n \in \mathcal{C}_i} w_j^{(n)} \qquad (3.39)$$

The number of clusters can be controlled by the parameter ε, the larger the neighborhood, the smaller the final number of particles. The exact product can then be calculated as described in Sec. 2.3.2.

3.8.1.2 Product of Gaussian Mixtures Approximation Using Nearest Neighbor Search

The following efficient approximation of the product reduces the combinatorial complexity from $\mathcal{O}(L^d)$ to $\mathcal{O}(Ld)$. Instead of multiplying each Gaussian with all Gaussians of the other mixtures, we multiply each Gaussian just with the most probable Gaussians. This is especially a valid assumption in the case where the bandwidth is small, since the remaining Gaussian products have a small contribution to the final density estimate. We can thus estimate the belief:

$$b_i(x_i) = \psi_j(x_j, y) \prod_{k \in \Upsilon(i)} m_{kj}(x_i)$$

$$= \psi_j(x_j, y) \prod_{k \in \Upsilon(i)} \sum_{n=1}^{N_k} w_{ki}^{(n)} \mathcal{N}\left(x_i; x_{ki}^{(n)}, \Lambda_{ki}\right)$$

$$\approx \psi_j(x_j, y) \sum_{k \in \Upsilon(i)} \sum_{n=1}^{N_k} w_{ki}^{(n)} \mathcal{N}\left(x_i; x_{ki}^{(n)}, \Lambda_{ki}\right) \prod_{j \in \Upsilon(i) \setminus k} w_{ji}^{(m_{\max})} \mathcal{N}\left(x_i; x_{ji}^{(m_{\max})}, \Lambda_{ji}\right)$$

$$(3.40)$$

with

$$m_{\max} = \underset{m=1,\ldots,N_j}{\arg\max} \ \mathcal{N}\left(x_{ji}^{(m)}; x_{ki}^{(n)}, \Lambda_{ki}\right) \tag{3.41}$$

Furthermore, if the covariance Λ_{ji} is constant we can also use a simple nearest neighbor search $m_{\max} = \arg\min_{m=1,\ldots,N_j} \|x_{ji}^{(m)} - x_{ki}^{(n)}\|$ allowing us to further speed up the calculation. Especially, if search structures like a kd-tree are used, fast calculations are possible.

3.8.2 Bottom-Up and Top-Down Message Passing

The recognition performance as well as the execution time of the inference procedure depend crucially on the number of particles. The more particles are used, the more accurate are the density distributions but the more time is spent for calculation. During bottom-up message passing, each node sends messages to all parent nodes, where the nodes are processed level by level. This processing can be seen as a breadth-first search, since the entire level with all nodes is exhaustively processed. Because we are in general interested in fast detection of high-level objects, the exhaustive search of all possible low-level hypotheses is often less important. We propose therefore to combine the bottom-up message passing with a depth-first search. The idea is trying to pass good hypotheses through the hierarchy, find good high-level hypotheses and omit bad hypotheses at the beginning. We perform bottom-up message passing in several sweeps. In each sweep a single particle or a subset of the particles is chosen according to a *selection rule*. We combine two different attributes to select a particle. First, we choose a node j according to its importance weight μ_j

$$j \sim \mu_j \tag{3.42}$$

This weight summarizes the amount of information the node contributes to the root nodes.

Definition 14 (Importance Weight). *The importance weight μ_j of node j is defined by the number of messages the root nodes receive directly or indirectly from node j during the bottom-up message passing.*

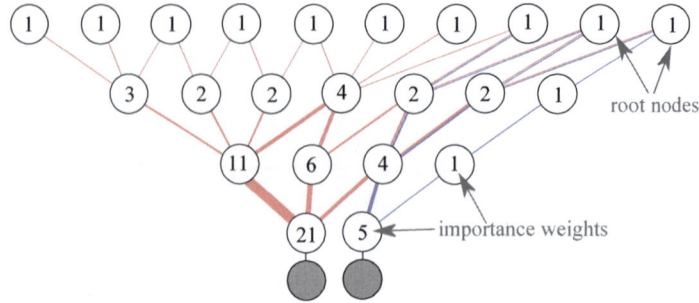

Fig. 3.18 Example of importance weights in a sharing structure: the left low-level node has an importance weight of 21 (red) and the right one an importance weight of 5 (blue). It is therefore more efficient to initiate bottom-up message passing sweeps based on the left low-level node.

It is recursively given by the weights of the parent nodes

$$\mu_j = \sum_{i \in \Gamma(j)} \mu_i \tag{3.43}$$

with $\mu_i = 1$ for all root nodes. Fig. 3.18 shows an example, where the left low-level node has an importance weight of 21 (red) while the right one has an importance weight of 5 (blue). Nodes with low importance weights represent in general object-specific parts and features, which are not shared. They contribute to one or just a few root nodes. Their evaluation is therefore costly and should be minimized in an efficient recognition approach. After node j is given, a particle $(x_j^{(n)}, w_j^{(n)})$ of the nonparametric density $b(x_j)$ is chosen according to its weight

$$\tilde{x}_j^{(n)} \sim w_j^{(n)} \tag{3.44}$$

and passed through the graph. At the beginning, particles are mainly chosen from low-level nodes. After a few sweeps particles at higher levels have been generated, so that it is also reasonable to initiate the message passing from higher levels. During bottom-up propagation, particles are sent to parent nodes, which are again chosen according to the importance weight $i \sim \mu_i$ with $i \in \Gamma(j)$.

The message passing is complicated by the fact that as single particles are pushed through the graph, the product calculation becomes challenging. This is because it is not guaranteed that all messages are already received and furthermore it is not guaranteed that the messages are complete, i.e. appropriate particles from the neighbors were already received. We will therefore often multiply with zero and thus destroy valid hypotheses. To solve this problem we introduce an additional top-down search step. If the messages from the other neighbors are empty or contain no appropriate particles, we send new messages downwards "searching" for low-level particles, that confirm the high-level hypothesis. The messages are processed top-down until appropriate particles are found or the lowest level is reached.

Let us now have a closer look at this top-down search. Assume node x_k sends a message $m_{ki}(x_i)$ containing a single particle $x_{ki}^{(1)}$ to its parent node x_i, which can now estimate the belief $b_i(x_i)$ by calculating the product from all incoming messages. In order to decrease the computational costs we approximate this product by searching for particles $x_{ji}^{(m_{max})}$, which are similar to $x_{ki}^{(1)}$ as described in Sec. 3.8.1.2. We accept particles $x_{ji}^{(m_{max})}$ with an acceptance rate

$$A\left(x_{ji}^{(m_{max})}\right) = \exp\left(-\frac{1}{2}\left(x_{ki}^{(1)} - x_{ji}^{(m_{max})}\right)^T \Lambda_{ki}^{-1}\left(x_{ki}^{(1)} - x_{ji}^{(m_{max})}\right)\right) \quad (3.45)$$

where Λ_{ki} denotes the covariance matrix associated to the potential between node k and i. If no particle was found or the particle was not accepted, a top-down message $m_{ij}(x_j)$ is sent from parent node x_i to its child node x_j. This message contains again just one particle $x_{ij}^{(1)}$. At node x_i we search for appropriate particles in the incoming messages from the children x_h with $h \in \Xi(j)$ according to eq. 3.41. Particles $x_{hj}^{(m_{max})}$ from the neighbors are accepted using the acceptance rate

$$A\left(x_{hj}^{(m_{max})}\right) = \begin{cases} \exp\left(-\frac{1}{2}\left(x_{ij}^{(1)} - x_{hj}^{(m_{max})}\right)^T \Lambda_{ij}^{-1}\left(x_{ij}^{(1)} - x_{hj}^{(m_{max})}\right)\right) & \ell_{x_i} \neq 1 \\ 1 & \text{otherwise} \end{cases}$$
$$(3.46)$$

where we separately handle nodes at level 1, which will always be accepted. This guarantees that we are able to detect objects despite occlusions as we will see later. If for each neighbor a particle is accepted, we can calculate the product of the Gaussian distributions, and add the result $x_j^{(1)}$ to the nonparametric representation of $b(x_j)$. The result can then be used to calculate a new message $m_{ji}(x_i)$ and subsequently to send it back to the parent node x_i, where it is multiplied with message $m_{ki}(x_i)$ and $m_{ji}(x_i)$. The process is recursively repeated for empty or unaccepted messages of each child. This message passing is illustrated in Fig. 3.19. While in Fig. 3.19(a) all particles are processed and sent upwards level by level, we process in Fig. 3.19(b) sequentially single particles, and pass them through the graph (the order is indicated by numbers). This process can be seen as an efficient bottom-up hypothesis generation step and a top-down hypothesis evaluation step.

This sampling scheme has some attractive properties. The nonparametric estimates of the beliefs contain no particles at the beginning. During the first sweep and in the top-down hypothesis evaluation step, we have therefore typically to step downwards until we reach the lowest level and find a valid hypothesis. However, since the estimates of the beliefs are "filled" with valid hypotheses in every sweep,

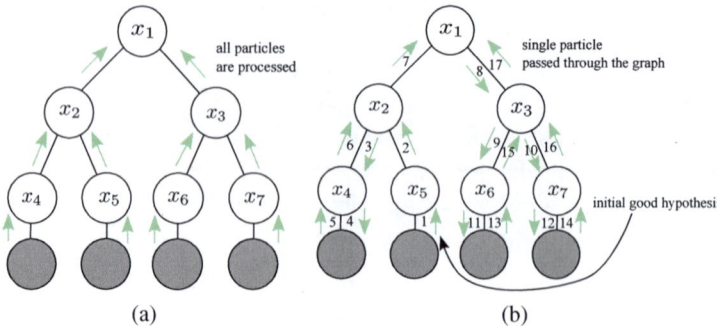

Fig. 3.19 Bottom-up and top-down message passing. a) Strict bottom-up message passing. b) Combined bottom-up and top-down message passing.

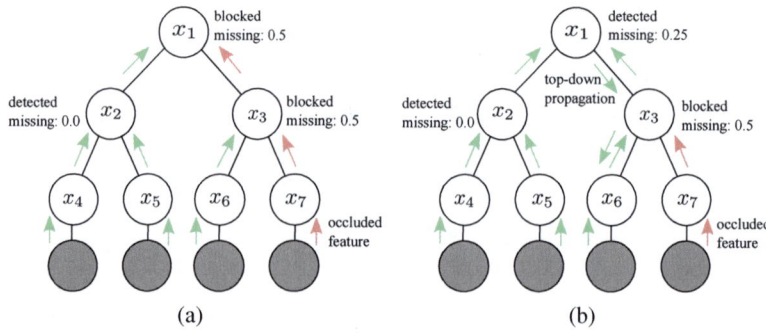

Fig. 3.20 Occlusion handling in hierarchical graphical models. a) The missing feature x_7 blocks the detection of x_3, and x_3, in turn, blocks x_1. b) The additional top-down step finds the present feature of x_6 and thus detects x_1 correctly.

it is likely that after a few sweeps valid high-level hypotheses are available, so that the top-down processing terminates after a few levels and does not have to be continued to the lowest level. Another advantage is that the approach processes and evaluates good hypotheses first and is thus able to find valid high-level hypotheses within a few sweeps. Especially, the top-down evaluation step increases the recognition performance since it induces a detailed search for local evidence in the context of parents. However, uncertain high-level hypothesis will in general still need more sweeps until they are found.

Another motivation for the top-down proceeding are occlusions. In Fig. 3.20 we show a simplified illustration of this situation. As can be seen in Fig. 3.20(a), missing low-level particles can block the detection of high-level parts. We assume that at least 60% of the low-level features have to be present in order to detect a new part. During simple bottom-up processing, the missing feature blocks the detection of

x_3, and x_3, in turn, blocks x_1. Compared with that, the top-down step allows to find the present feature of x_6 and thus detects x_1 correctly (see Fig. 3.20(b)). In a strict bottom-up processing we therefore have to process a large number of hypotheses in order to find occluded or unlikely parts. On the other hand, the combined bottom-up and top-down processing allows to deal with occlusions since the top-down evaluation steps explicitly searches for occluded or unlikely low-level hypotheses.

Chapter 4
Learning of Hierarchical Models

Given some kind of input data, the aim of learning is to build models that are able to represent the input data and generalize them for the recognition of previously unseen data. The input data is typically encoded in a training set that contains images or image sequences for different classes. Many different training datasets are publicly available. The caltech-256 dataset [83] is a challenging set of 256 object categories containing a total of 30607 images (see Fig. 4.1(a)). It was collected by choosing a set of object categories and downloading appropriate examples from the internet with a minimum number of 80 images in each category. Thus, the object images are captured from different view-points, contain background clutter, different object appearances and different scales. Similar datasets have also been collected for activity recognition. The Weizmann dataset [81] contains a total of 90 videos containing 10 action categories performed by 9 people. The actions were captured in front of a simple background and from one fixed camera position (see Fig. 4.1(b)). A more challenging activity dataset is the Hollywood Human Actions dataset [143] whose training samples have a large variability of scale, viewpoint and background. It is a collection of video sequences from hollywood movies and contains realistic samples of human actions like kissing, answering a phone or getting out of a car (see Fig. 4.1(c)).

Different from these training datasets are those used in this monograph. Our motivation is to learn a robust representation of an object or an activity from few or even single training instances. The dataset should therefore not contain a large random collection of object instances, but should instead cover the whole possible configuration or appearance space. The Columbia Object Image Library (COIL-100) [165] is a collection of color images of 100 objects (see Fig. 4.1(d)). The objects were rotated on a turntable through 360 degrees to vary object pose. During rotation of each object, images were taken at pose intervals of 5 degrees leading to 72 poses per object. Due to the missing background clutter and occlusions, the COIL database seems less challenging. However, the viewpoint changes induce large variations of the appearance as well as of the geometrical arrangement of the parts. This is an ideal setting for our proposed hierarchical framework and will be used in Chap. 5. Another dataset is the human pose dataset (see Fig. 4.1(e)), which is used in Chap.

© Springer International Publishing Switzerland 2015   67
J. Spehr, *On Hierarchical Models for Visual Recognition & Learning of Objects, Scenes, & Activities,*
Studies in Systems, Decision and Control 11, DOI: 10.1007/978-3-319-11325-8_4

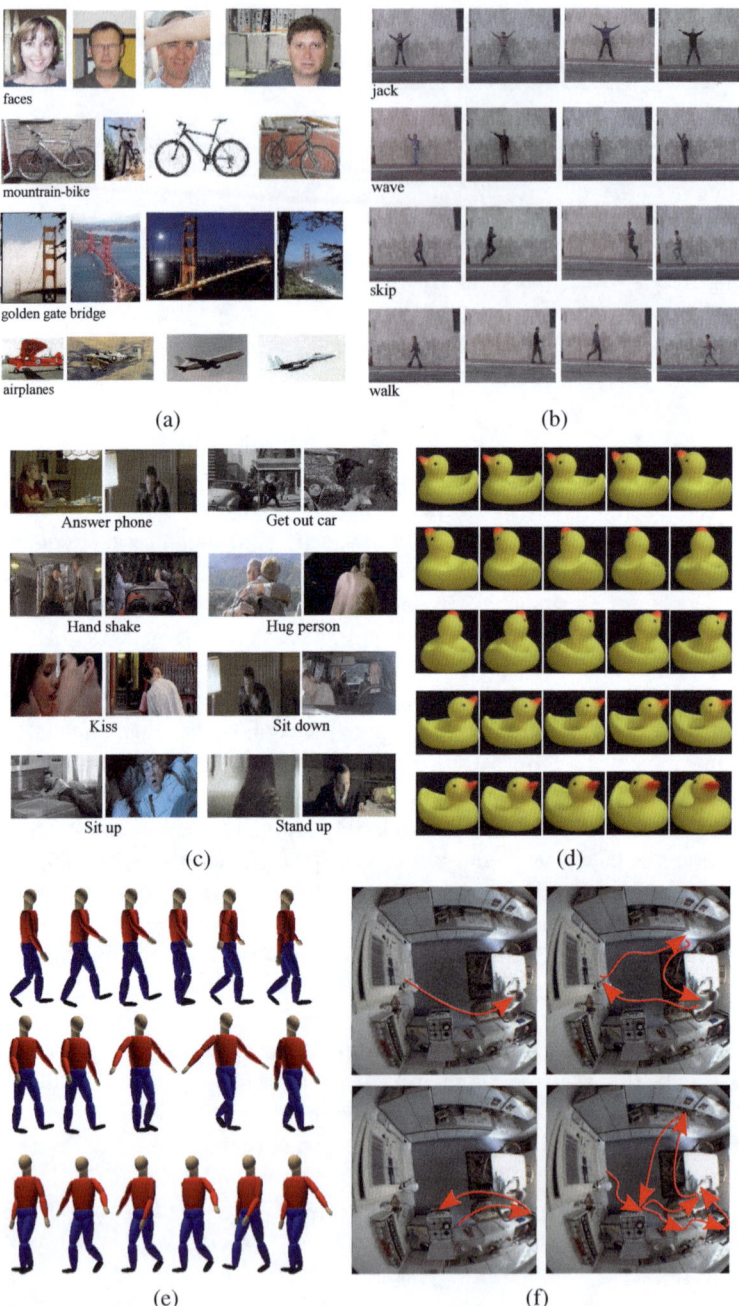

Fig. 4.1 Comparison of common datasets with the datasets used in this monograph. a) Caltech-256 [83]. b) Weizmann [81]. c) Hollywood Human Actions [143]. d) Coil-100 [165] (this monograph). e) Human pose dataset (this monograph). f) Human behavior dataset (this monograph).

6 for the purpose of human gait analysis. It is a collection of different human poses, showing a walking person in different configurations. We generated the dataset using a 3d model of the human body and render images of different configurations and viewpoints. Human behavior patterns are analyzed in Chap. 7. The corresponding dataset contains a set of top view video sequences where each sequence represents the daily routines of a person in its home environment (see Fig. 4.1(f)). These daily routines include activities of daily living like preparing food, changing rooms, answering phones, and so on. Due to the high variations of these activities and their changing spatial and temporal context the representation is challenging. The aim of our learning framework is to get a valid representation of the daily routines even after one day, so that, the next day, we can recognize activities of the previous day and also detect new unseen activities. This can be used to detect anomalies, to learn new unseen motion patterns and to adapt our model.

Learning of hierarchical models involves two steps: structure learning and parameter learning. Structure learning is a computationally challenging problem due to the exponentially large number of possible structures, that underly a set of variables. Since the whole search space over all possible structures cannot be searched, we must rely on constraints and heuristics to guide the search. Whilst in general structure learning is intractable, appropriate constraints are available for hierarchical structures.

In Sec. 4.1 we will review related work from the literature and in Sec. 4.2 we introduce our new learning framework.

4.1 Related Work

Most learning approaches are dedicated to specific models or classifiers like linear regression, Bayesian networks or support vector machines. Often, learning corresponds to a search for the best model. Bayesian model selection, for example, uses the rules of probability theory to find the best model [16, 10]. Given a training set D, we can express the posterior distribution for model M using Bayes'rule

$$p(M|D) \propto p(D|M)p(M) \tag{4.1}$$

Preference for different models is expressed by the prior $p(M)$. It can be used to incorporate additional information, which does not depend on the training set, into the learning, like e.g. heuristics. The model evidence is the probability of the data D given the model M, and is computed by integrating over the unknown parameter values θ in that model

$$p(D|M) = \int_\theta p(D|\theta)p(\theta|M)d\theta \tag{4.2}$$

Due to that, marginalization $p(D|M)$ is often also called marginal likelihood.

In the following we will briefly review learning approaches related to hierarchical compositions similar to our model. In Sec. 4.1.1 we summarize supervised learning approaches which use manually labeled training sets or predefined hierarchies. Un-supervised structure learning approaches are reviewed in Sec. 4.1.2 and divided into two categories (bottom-up and and top-down).

4.1.1 Supervised Structure Learning

In a supervised learning framework, the hierarchical structure is inferred from super-vised (labeled) training data. The training data can be seen as examples that consist of input objects and associated output values. The input objects are for example image patches of an object or a part that were manually labeled in an image and as-signed to a specific element of the hierarchy. Zhu and Mumford [269] used a learn-ing framework, where production rules as well as image segments were manually defined. They needed a full time annotation team for parsing the image structures. The annotated dataset was used to construct AND/OR graphs for object and scene categories.

One major class of supervised learning approaches uses predefined visual prim-itives to build the hierarchy [53]. Biederman [13] proposed a set of generalized cylinders for representing 3d object elements. Geman et al. [80] used a library that includes letters, and intermediate-level representations like lines, arcs, T-junctions and L-junctions.

Often also the hierarchical structure is manually defined by selecting an appropri-ate topology. Fergus et al. [58] used a simple two-level hierarchy with a predefined number of parts. Similarly, Bouchard and Triggs [19] used a predefined three-layer hierarchy, where the number of parts in each layer of the hierarchy was fixed by hand.

4.1.2 Unsupervised Structure Learning

To obtain a better overview of unsupervised structure learning approaches from the literature, we divide them into two classes concerning their learning direction: bottom-up (Sec. 4.1.2.1) and top-down (Sec. 4.1.2.2).

4.1.2.1 Bottom-Up Structure Learning

Bottom-up structure learning approaches rely generally on searching for correlated variables. Two variables and their associated features are correlated, if they are likely to co-occur in the same spatial neighborhood. The concept of co-occurrence analy-sis is a widely used method for the discovery of relations between variables. In data mining, the co-occurrence analysis is used to learn association rules. An example is the classic apriori algorithm [2], which efficiently finds frequent subsets by count-ing their occurrences in a tree structure. It proceeds "bottom-up" in a breadth-first search, extents frequent subsets one item at a time, and measures the support of the extented subset by testing it against the data. Another domain, where co-occurrence analysis is applied, is the text document analysis. Wettler and Rapp [247] proposed a statistical model which predicts the strengths of word-associations from the relative frequencies of the common occurrences of words. Similarily, Morita et al. [157] described the co-occurrence word information in the natural language processing system.

There is also experimental psychological evidence for the importance of co-occurences. Fiser and Aslin [70, 69] investigated in three experiments the ability of human observers to extract the joint and conditional probabilities of shape co-occurrences during passive viewing of complex visual scenes. They found that subjects learned statistics from the spatial arrangement of shapes and concluded that this supports Barlow's theory [11] of visual recognition, which states that "suspicious coincidences" of elements during recognition is necessary for efficient learning of new visual features.

In visual recognition, the idea of co-occurence learning was also used for learning of visual hierarchies [205, 268, 66]. Scalzo and Piater [205] used co-occurence learning to first determine the hierarchical latent structure by searching for spatial correlations, and then estimated the model parameters like the spatial relations. During spatial correlation extraction, each feature pair $[f_i, f_j]$, which is observed in a neighborhood, votes for the corresponding observation $[f_i, f_j, p_r]$ in the voting table $\mathcal{T}$, where $p_r \in R^2$ denotes the relative position between f_i and f_j. Pairs $[f_i, f_j]$ are considered to be spatially correlated if $\sum_{p_r} \mathcal{T}[f_i, f_j, p_r] > t_c$. During spatial relation estimation, they used the EM algorithm to estimate the parameters of the spatial relation between each correlated feature pair $[f_i, f_j] \in \mathcal{T}$. Finally, a new feature on the next higher level was generated, when a reliable reciprocal spatial correlation was detected between two features $[f_i, f_j]$. The new feature represents a composition of the two subfeatures $[f_i, f_j]$ located at the midpoint between the subfeatures. While the number of layers is principally unrestricted, one major drawback of the approach is the restriction to at most two children. Fidler and Leonardis [66] extended this learning framework to an arbitrary number of children. They found correlated feature pairs, as before, and extended them iteratively by extracting correlations between the composed features and other single features.

One problem of the bottom-up processing is that subparts may overlap and thus may lead to an inefficient representation. Furthermore, different feature compositions may overlap and thus may represent the same object leading to a large number of possible structures. In order to reduce these undesirable properties Fidler and Leonardis [66] introduced *local inhibition* to constrain the maximum ratio of overlapping features. Zhu et al. [268] introduced similar approaches to confine the number of proposals to a practical number and avoid an exponential increase of possible structures. Their *suspicious coincidence* principle removes concepts which occur infrequently and their *competitive exclusion* principle removes concepts whose instances overlap with those of other concepts. One of the main drawbacks of bottom-up learning approaches is, that they need a large dataset of training images in order to find statistically significant correlations. For example, Fidler and Leonardis [66] applied their method to a collection of 3,200 images containing just 15 categories. The large training sets were necessary in order to find spatial relations between features and initiate the bottom-up process. Smaller training sets are generally leading to incomplete object representations where shape variations or articulated parts are missing. For the datasets used in this monograph, where generally just one instance of an object or object view is given, these approaches are therefore inappropriate.

As a workaround, Zhu et al. [268] proposed to combine the bottom-up process with a top-down "completion" process. In this top-down process, they filled in the missing parts of the hierarchy and added a dense representation at the lowest level. For that, they used a greedy strategy, which examines every node in the hierarchy and seeks to add a substructure from the dictionary.

4.1.2.2   Top-Down Structure Learning

Top-down structure learning approaches start at the top, where high-level objects are represented, and iteratively decompose objects and parts into smaller parts. As far as we know, top-down structure learning for compositional hierarchies was so far only proposed by Ephstein and Ullman [48, 49]. They introduced a top-down method for automatically learning their visual hierarchies, which are similar to the HMAX model [192]. Their approach starts at the top level, where informative fragments are extracted from the training set. The fragments are then decomposed into object parts using an optimal decomposition criterion. The criterion evaluates candidate fragments f_i by the amount of mutual information [33] they deliver about a class C. Mutual information is a quantity that measures the mutual dependence of two random variables

$$MI(X;Y) = \sum_{x \in X} \sum_{y \in Y} p(x,y) \log \left(\frac{p(x,y)}{p(x)p(y)} \right) \qquad (4.3)$$

$MI(X;Y) = 0$ means the variables are independent. The mutual information $MI(f_i;C)$ is therefore the amount by which the knowledge provided by the candidate fragment f_i decreases the uncertainty of the class C. The algorithm finds the fragment with the highest mutual information score, and after that identifies the next fragment that delivers the maximal amount of additional information with respect to previously selected fragments. This selection process is iterated downwards for each layer. Finally, the algorithm terminates with simple low level features which cannot be usefully decomposed into simpler features further (see [48, 49] for a detailed overview). The structure learning builds hierarchies with typically three layers and five children on average, and was applied to faces, cars, horses, cows and airplanes.

Although the approach uses a similar top-down learning as our approach, the methods differs crucially in learning objectives and model assumptions. The main difference is that Ephstein and Ullman built hierarchies that maximize mutual information about a class; in contrast to this, our approach builds hierarchies that maximize the reusability of parts. In order to achieve this maximal reusability we allow very flexible hierarchies which are not restricted to a specific number of levels, number of children, or a specific type of interlayer connections. Furthermore, Ephstein and Ullman needed about 200 training samples per class to build their hierarchy. This is another important difference to our approach since we are considering datasets, where even single training samples are learned.

4.2 Hierarchical Structure Learning as an Optimization Problem

The aim of our learning framework is to build a hierarchical representation for each training instance. Our approach is therefore more related to *instance-based learning* (also called memory-based learning) approaches [3] which represent a model by examples (instances). This model is one of the fundamental concepts in human learning theory [188, 25]. It constructs models directly from the training instances themselves leading to a representation that grows with the data. Instance-based learning is a kind of lazy learning since generalization is delayed until a new instance must be classified. This is different from our approach, where new instances are incorporated into hierarchies, which share structure with other instances. The hierarchical representation thus guarantees a robust representation that is able to generalize new data on different abstraction layers. Furthermore, the sharing guarantees that the complexity of the inference does not grow linearly with the number of instances. In a large set of hierarchies, we would ideally suspect that the parts of a new instance are already contained in the set of hierarchies, so that the additional effort of incorporating this new instance is low. Maximal reusability of parts can be allowed only if the hierarchical structure is not restricted to a certain number of levels or children. Another important property is that the children of a parent must not be element of the same level and especially not on the next lower level. This restriction would lead to inaccurate and non optimal representations since asymmetric objects have generally to be decomposed into parts of different complexities, and hence, different hierarchy levels.

In the following we will introduce our hierarchical learning framework. We will distinguish between two different learning types. In the first, we assume we have a set of n observations given

$$\mathcal{O} = \{\mathcal{O}_1, ..., \mathcal{O}_n\} \tag{4.4}$$

The aim is to learn a hierarchical representation for each observation $\mathcal{O}_i$. Since in this case all observations are present during learning, this type is called offline learning (Sec. 4.2.1). The other type assumes that an incomplete set of instances is given and extended one at a time

$$\mathcal{O}^t = \mathcal{O}^{t-1} \cup \{\mathcal{O}_n\} \tag{4.5}$$

$$\mathcal{O}^1 = \{\mathcal{O}_1\} \tag{4.6}$$

This type of learning is referred to as online learning which will be introduced in Sec. 4.2.2.

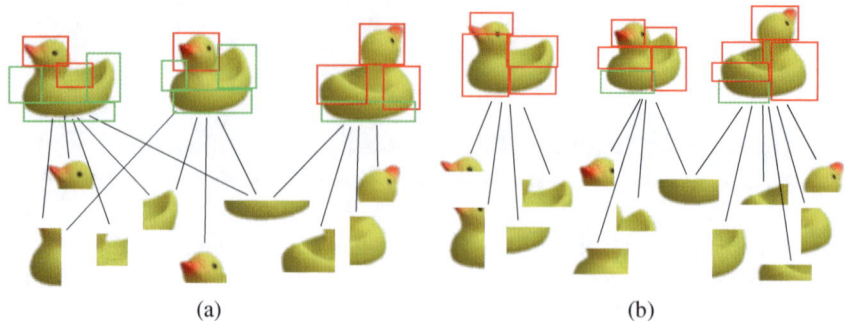

Fig. 4.2 Random decomposition of three object instances into parts with different degrees of sharing. (a) High degree of sharing (shared parts highlighted in green). (b) Low degree of sharing.

4.2.1 Offline Learning

We now introduce our unsupervised learning approach, which builds efficient compositional hierarchies based on a set of training instances. The model is represented by a set of hierarchies $\mathcal{G} = \{\mathcal{G}_1, ..., \mathcal{G}_n\}$, which are undirected tree-structured graphs $\mathcal{G}_i = (\mathcal{V}_i, \mathcal{E}_i)$, and a set of additional edges $\mathcal{L}$, which represent links between the different graphs.

First, an appropriate low level detector as discussed in Sec. 3.5.1 is applied to the training data, which delivers a set

$$\mathcal{O} = \{\mathcal{O}_1, ..., \mathcal{O}_n\} \tag{4.7}$$

of n training instances. Each feature set $\mathcal{O}_i = \{o_1, ..., o_n\}$ contains the detected features in the ith training instance. A feature is defined as a pair $o = (\mu, \delta)$ with a position $\mu \in \mathbb{R}^d$, and a descriptor δ that describes the local appearance. The learning starts by associating each feature set $\mathcal{O}_i$ with a new root node of graph $\mathcal{G}_i$ at hierarchy level ℓ according to the level selection suggested in Sec. 3.7. The feature set of root node x_i thus ideally corresponds to the training instance

$$\mathcal{F}_i = \mathcal{O}_i \tag{4.8}$$

During learning the feature sets $\mathcal{O} = \{\mathcal{O}_1, ..., \mathcal{O}_n\}$ are iteratively decomposed into smaller subsets and the aim of the learning process is to find an efficient structure, where as many of the subsets as possible are shared between the different object parts.

We learn the structure, i.e. the set of graphs $\mathcal{G}$ and the parameters $\theta_{\mathcal{G}}$, using a maximum a posteriori (MAP) estimation

$$(\hat{\mathcal{G}}, \hat{\theta}_{\mathcal{G}}) = \operatorname{argmax}_{\mathcal{G}, \theta_{\mathcal{G}}} p(\mathcal{G}, \theta_{\mathcal{G}} | \mathcal{O}) \tag{4.9}$$

$$= \operatorname{argmax}_{\mathcal{G}, \theta_{\mathcal{G}}} \left\{ \log p(\mathcal{G}, \theta_{\mathcal{G}}) + \sum_{i=1}^{n} \log p(\mathcal{O}_i | \mathcal{G}_i, \theta_{\mathcal{G}_i}) \right\} \tag{4.10}$$

where the prior $p(\mathcal{G}, \theta_{\mathcal{G}})$ is used to enforce preference to graphs with many shared nodes and thus prefer more efficient hierarchies. Here, we set

$$p(\mathcal{G}, \theta_{\mathcal{G}}) \propto \exp\left\{ -\gamma_1 |\mathcal{V}_{\mathcal{G}}| - \gamma_2 |\mathcal{E}_{\mathcal{G}}| \right\} \tag{4.11}$$

where $|\mathcal{V}_{\mathcal{G}}|$ is the number of nodes and $|\mathcal{E}_{\mathcal{G}}|$ is the number of edges in the sharing structure of the graph. The parameters γ_1 and γ_2 are used to weight the influence on the learning. The likelihood $p(\mathcal{O}_i | \mathcal{G}_i, \theta_{\mathcal{G}_i})$ represents the probability that training instance $\mathcal{O}_i$ is generated by a graph $(\mathcal{G}_i, \theta_{\mathcal{G}_i})$ and adjusts the graph structure to the training data

$$p(\mathcal{O}_i | \mathcal{G}, \theta_{\mathcal{G}}) \propto \exp\left\{ -\gamma_d dist(\mathcal{F}_i, \mathcal{O}_i) \right\} \tag{4.12}$$

Due to the many unknowns an exact optimization is not feasible. Therefore, we use a Metropolis-Hasting sampling scheme in order to obtain a representative set of samples of the distribution $p(\mathcal{G}, \theta_{\mathcal{G}} | \mathcal{O})$. The MAP result $(\hat{\mathcal{G}}, \hat{\theta}_{\mathcal{G}})$ corresponds to the sample that maximizes $p(\mathcal{G}, \theta_{\mathcal{G}} | \mathcal{O})$. Metropolis-Hasting [88, 16] is a Markov chain Monte Carlo sampling method, which allows to generate samples from a probability distribution for which direct sampling is difficult, like e.g. in our case due to high dimensionality of the sampling space. The algorithm uses a current state $(\mathcal{G}^{(\tau)}, \theta_{\mathcal{G}}^{(\tau)})$, which represents a set of hierarchies and their parameters, at step τ and draws a sample $(\mathcal{G}^*, \theta_{\mathcal{G}}^*)$ from the proposal distribution $q((\mathcal{G}^*, \theta_{\mathcal{G}}^*) | (\mathcal{G}^{(\tau)}, \theta_{\mathcal{G}}^{(\tau)}))$. The new sample is accepted with probability

$$A((\mathcal{G}^*, \theta_{\mathcal{G}}^*), (\mathcal{G}^{(\tau)}, \theta_{\mathcal{G}}^{(\tau)})) = \min\left(1, \frac{\tilde{p}(\mathcal{G}^*, \theta_{\mathcal{G}}^* | \mathcal{O}) q((\mathcal{G}^{(\tau)}, \theta_{\mathcal{G}}^{(\tau)}) | (\mathcal{G}^*, \theta_{\mathcal{G}}^*))}{\tilde{p}(\mathcal{G}^{(\tau)}, \theta_{\mathcal{G}}^{(\tau)} | \mathcal{O}) q((\mathcal{G}^*, \theta_{\mathcal{G}}^*) | (\mathcal{G}^{(\tau)}, \theta_{\mathcal{G}}^{(\tau)}))} \right) \tag{4.13}$$

where $\tilde{p}(\cdot)$ is the unnormalized posteriori distribution $\tilde{p}(\cdot) = Z_p p(\cdot)$ with the normalizing constant Z_p.

Unfortunately, the right choice of an efficient proposal distribution is quite challenging since each decomposition depends on the decomposition of the higher levels. We, therefore, propose to combine the sampling with a top-down decomposition of the root nodes. We estimate the posterior separately for each level and sample from the whole hierarchy in several top-down sampling iterations. Starting at the highest level ℓ_{top}, the sampling processes all root nodes and their associated features sets of this level $\mathcal{O}^{\ell_{top}} = \{\mathcal{O}_i | \ell_i = \ell_{top}\}$. An auxiliary set $\mathcal{A} = \{\mathcal{A}^{\ell_1}, ..., \mathcal{A}^{\ell_{top}}\}$ is used to store the feature sets of each layer. Initially, the auxiliary feature set $\mathcal{A}^{\ell_i} = \mathcal{O}^{\ell_i}$ contains all root nodes of ℓ_i. The feature sets are randomly decomposed into subsets, which are associated to a level according to the number of features, see eq. 3.33. Please note, that this association step guarantees a flexible hierarchical structure since it includes no interlayer constraints. This decomposition represents a partial structure of our set of hierarchies $(\mathcal{G}^{(\tau),\ell_{top}}, \theta_{\mathcal{G}}^{(\tau),\ell_{top}})$ and can be used to calculate the posterior $p(\mathcal{G}^{(\tau),\ell_{top}}, \theta_{\mathcal{G}}^{(\tau),\ell_{top}}|\mathcal{O})$. For this, we have to determine the number of nodes that are shared and count the number of nodes and edges of $\mathcal{G}^{(\tau),\ell_{top}}$ according to eq. 4.11 . We use a density based clustering algorithm similar to that described in Sec. 3.8.1.1 in order to determine sharing groups, where all nodes fulfill the sharing criterion among each other (see Sec. 3.7.3). The number of edges and nodes of the new sharing structure is used to calculate the probability of the prior. One consequence of the sharing is, that the feature set $\mathcal{F}_i$ of root node x_i changes since the shared parts are replaced by the sharing representative. To avoid too large differences, the likelihood (see eq. 4.12) adapts the feature set to the observation $\mathcal{A}_i$. After we have calculated the posterior we can use the proposal function in order to sample a new decomposition $(\mathcal{G}^{*,\ell_{top}}, \theta_{\mathcal{G}}^{*,\ell_{top}})$.

We associate a state vectors z to each feature set $\mathcal{A}_i$ in order to simplify the sampling and facilitate the formulation of appropriate proposal functions. The hierarchy is thus formulated as a function of the state vector

$$(\mathcal{G}, \theta_{\mathcal{G}}) = (\mathcal{G}(z), \theta_{\mathcal{G}}(z)) \tag{4.14}$$

where $z = \{z_1, ..., z_{\ell_{top}}\}$ is the set of all state vectors with $z_\ell = \{z_i|\ell_i = \ell\}$. The state vector is used to guide the decomposition and to restrict the set of possible decompositions to reasonable ones. While the proposal function could in principle propose arbitrary decompositions, often just a small subset is actually reasonable (see Fig. 4.3). A state vector could thus be used to restrict the decomposition by regarding e.g. spatial or appearance information. In the following chapters we will use symmetric proposal distributions $q(z_\ell^*|z_\ell^{(\tau)})$, so that the sampling corresponds to Metropolis sampling with an acceptance rate for a new candidate

$$A(z^*, z^{(\tau)}) = \min\left(1, \frac{\tilde{p}(\mathcal{G}(z^*), \theta_{\mathcal{G}}(z^*)|\mathcal{O})}{\tilde{p}(\mathcal{G}(z^{(\tau)}), \theta_{\mathcal{G}}(z^{(\tau)})|\mathcal{O})}\right) \tag{4.15}$$

After a burning-in period of 20 iterations on average the sampling reaches a likely decomposition of the highest level, where the initial state is "forgotten". The samples, which are generated during the burn-in period, are discarded. We terminate the sampling and add each new part to the auxiliary set $\mathcal{A}^{\ell_i}$ according to the appropriate level ℓ_i. Here, shared parts are just added once, by means of the representative set $\mathcal{F}_r$, leading to a simplified and reduced feature set. After this, the algorithm starts the same procedure at the next lower level $\ell_{top} - 1$. As before, this level contains root nodes but also parts determined in the previous decomposition. The sampling is iterated until level one is reached, and the top-down sampling terminates. After one top-down decomposition sampling, we have one sample of the whole set of hierarchies. An example of the top-down decomposition is illustrated in Fig. 4.4. The routine is repeated several times to get a set of samples.

Fig. 4.3 Segmentation examples: a) Unreasonable segmentation into two parts. b) Spatial segmentation into two parts. c) Spatial segmentation into four parts. d) Appearance segmentation into three parts.

The final result $(\hat{\mathcal{G}}, \hat{\theta}_{\mathcal{G}})$ is selected to be the sample that maximizes $p(\mathcal{G}, \theta_{\mathcal{G}} | \mathcal{O})$. We build the hierarchical structure using the auxiliary set $\mathcal{A} = \{\mathcal{A}^{\ell_1}, ..., \mathcal{A}^{\ell_{top}}\}$. Each element of $\mathcal{A}$ corresponds to a node x_i and its associated feature set

$$\mathcal{F}_i = \mathcal{A}_i \qquad (4.16)$$

The node x_i is added to $\mathcal{V}$ and edges $e \in \mathcal{E}$ are defined between x_i and its children, which are given by the decomposition structure. Additionally, sharing links $e \in \mathcal{S}$ are established based on the shared structure determined during clustering. The whole algorithm is summarized in Alg. 1 and the top-down decomposition in Alg. 2.

4.2.2 Online Learning

Online learning allows to adapt the model to previously unseen data. Since our representation is instance-based, we can simply adapt the model by adding new instances. Another motivation of online learning is that the proposed offline learning needs to calculate a distance matrix in order to determine the sharing structure. Unfortunately, this distance matrix has to be calculated in every evaluation step and is one of the computationally most expensive operations. Online learning reduces the complexity of this operation und corresponds to a sequential learning. Although the learning optimizes the hierarchical representation based on an incomplete subset of the learning instance, the results achieve a similar efficient structure as completely offline learned models.

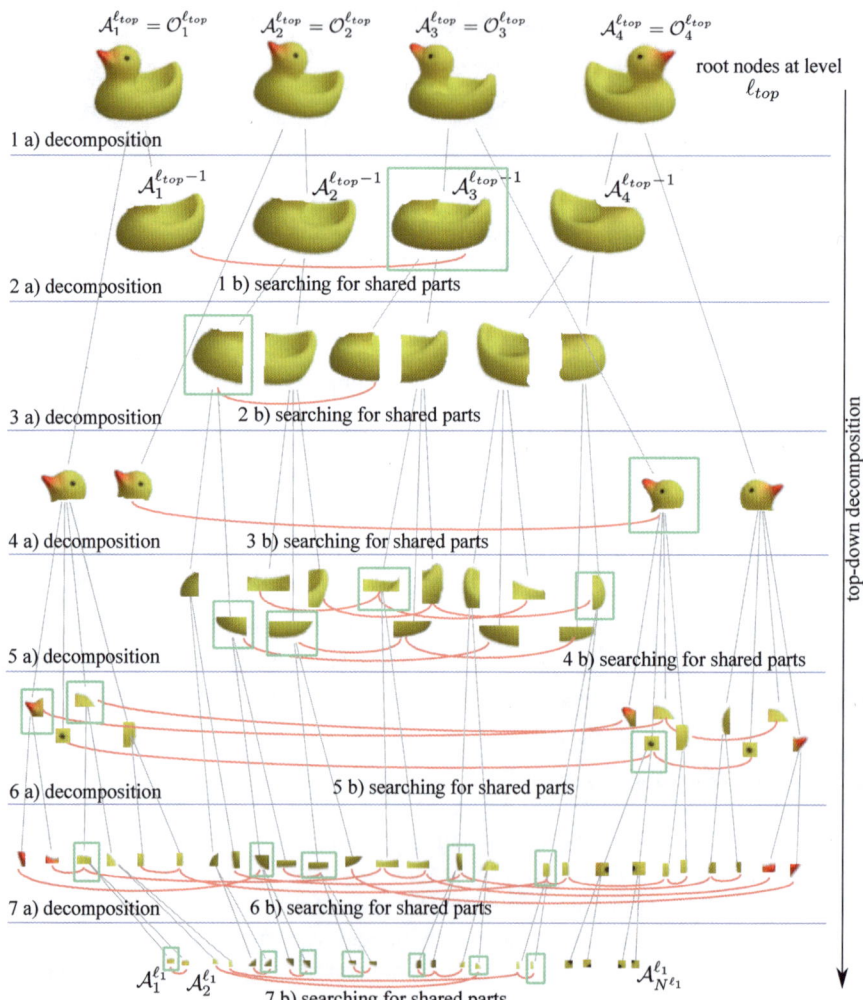

Fig. 4.4 Illustration of the top-down decomposition. The initial set of root nodes is randomly decomposed into parts. The parts are associated to hierarchy levels according to their number of low-level features. Then, similar parts are searched, which can be shared. The representatives of a sharing group are highlighted by a green bounding box. After that, the parts are again decomposed, and so on. Please note, that after shared parts are found, just the representative is further decomposed.

OFFLINE LEARNING($\mathcal{O}_1, ..., \mathcal{O}_n$)

1. Initialize the auxiliary feature set $\mathcal{A}$ and the set of state vectors z

    a. Associate each feature set $\mathcal{O}_i$ with a level ℓ_i according to eq. 3.33.
    b. Add each feature set $\mathcal{O}_i$ to the auxiliary feature set $\mathcal{A}^{\ell_i}$.
    c. Add for each feature set $\mathcal{O}_i$ a randomly initiated state vector to $z^{(\tau)}$.

2. Use the proposal distribution $q(z^*|z^{(\tau)})$ to get a new state vector z^*.
3. TOP-DOWN DECOMPOSITION($z^*, \mathcal{A}^{\ell_1}, ..., \mathcal{A}^{\ell_{top}}$).
4. Accept the new set of state vectors z^* with the posterior $p(\mathcal{G}(z^*), \theta_\mathcal{G}(z^*)|\mathcal{O})$ according to the acceptance rate given in eq. 4.15.
5. Repeat steps 1-4 (skip step 1c) for T_{off} iterations.
6. Build the hierarchical structure using the sample that maximizes $p(\mathcal{G}, \theta_\mathcal{G}|\mathcal{O})$.

    a. Each element of $\mathcal{A}$ corresponds to a node x_i that is added to $\mathcal{V}$.
    b. The edges $e \in \mathcal{E}$ are defined between x_i and its children, which are given by the decomposition structure.
    c. Sharing links $e \in \mathcal{S}$ are established based on the shared structure and the determined representatives.

Algorithm 1 Offline learning: Given a set of training instances $\mathcal{O}_1, ..., \mathcal{O}_n$, offline learning builds for each instance a hierarchy using the Metropolis sampler. During the sampling a top-down decomposition and a set of state vectors are used to guide the decomposition and efficiently find valid hierarchy hypotheses.

TOP-DOWN DECOMPOSITION($z^\tau, \mathcal{A}^{\ell_1}, ..., \mathcal{A}^{\ell_{top}}$)
for $\ell \leftarrow \ell_{top}$ to ℓ_1 **do**

1.   Use the proposal distributions $q(z_\ell^*|z_\ell^{(\tau)})$ to get a new state vector z_ℓ^*.
2.   Decompose the feature sets of $\mathcal{A}^\ell$ into subsets according to z_ℓ^*.
3.   Calculate the posterior $p(\mathcal{G}^\ell(z_\ell^*), \theta_\mathcal{G}^\ell(z_\ell^*)|\mathcal{O})$.

    a.   Determine sharing groups using a clustering algorithm and calculate the representative of each sharing group according to eq. 3.37.
    b.   Estimate the prior $p(\mathcal{G}, \theta_\mathcal{G})$ using eq. 4.11 and the likelihood $p(\mathcal{O}_i|\mathcal{G}_i, \theta_{\mathcal{G}_i})$ using eq. 4.12.

4.   Accept the new decomposition with the acceptance rate given in eq. 4.15.
5.   Repeat steps 1-4 for T_{td} iterations.
6.   Update the feature sets $\mathcal{A}^{\ell_1}, ..., \mathcal{A}^{\ell_{top}}$ and the state vectors z^τ

    a.   Associate each new feature subset to a level ℓ_i according to eq. 3.33.
    b.   Add each new feature subset to the auxiliary feature set $\mathcal{A}^{\ell_i}$ (for sharing groups just the representative is added).
    c.   Add for each new feature subset a randomly initiated state vector to z^τ.

end for

Algorithm 2 The top-down decomposition is a subroutine within the Metropolis sampling scheme. It iteratively decomposes the initial feature sets into smaller subsets while maximizing the sharing between the feature sets.

We learn the structure, similar to the offline learning, using a maximum a posteriori (MAP) estimation

$$(\hat{\mathcal{G}}_n, \hat{\theta}_{\mathcal{G}_n}) = \text{argmax}_{\mathcal{G}_n, \theta_{\mathcal{G}_n}}\ p(\mathcal{G}_n, \theta_{\mathcal{G}_n} | \mathcal{O}_n, \mathcal{G}^{n-1}, \theta_{\mathcal{G}}^{n-1}) \quad\quad (4.17)$$

$$= \text{argmax}_{\mathcal{G}_n, \theta_{\mathcal{G}_n}}\ \{\log p(\mathcal{G}_n, \theta_{\mathcal{G}_n} | \mathcal{G}^{n-1}, \theta_{\mathcal{G}}^{n-1}) + \log p(\mathcal{O}_n | \mathcal{G}_n, \theta_{\mathcal{G}_n})\} \quad\quad (4.18)$$

where $p(\mathcal{G}_n, \theta_{\mathcal{G}_n} | \mathcal{O}_n, \mathcal{G}^{n-1}, \theta_{\mathcal{G}}^{n-1})$ is the prior of the hierarchical representation of $\mathcal{O}_n$ given the previously learned set of hierarchies $\mathcal{G}^{n-1}, \theta_{\mathcal{G}}^{n-1}$ (see eq. 4.11). The likelihood $p(\mathcal{O}_n | \mathcal{G}_n, \theta_{\mathcal{G}_n})$ is defined similar to eq. 4.12.

There are two reasonable sampling approaches. The first one, *top-down learning*, proceeds similar to the offline learning. The only difference is that the auxiliary set $\mathcal{A} = \{\mathcal{A}^{\ell_1}, ..., \mathcal{A}^{\ell_{top}}\}$ is initiated with the feature sets of the already learned hierarchy $\mathcal{A}^{\ell_i} = \mathcal{F}^{\ell_i}$. During learning, the root node is randomly decomposed into subsets and similar feature sets of the already learned hierarchy are searched (see Alg. 3).

One of the drawbacks of the first solution is the random decomposition. In the worst case, it can take many iterations until a good decomposition is found, i.e. high level parts are re-used. The second sampling approach guides the decomposition by means of the already learned hierarchy set. The idea is to use a "learning by recognition" framework in order to find good decompositions. During *top-down learning by recognition*, the hierarchy $\mathcal{G}^{n-1}, \theta_{\mathcal{G}}^{n-1}$ is used to detect parts in the training sample. These parts are then used to find reasonable decompositions. The decomposition proceeds as follows: It starts with the new feature set $\mathcal{O}_n$ and searches sequentially for good feature sets, which were detected. A detected feature sets $\mathcal{F}_i$ is chosen according to the number of overlapping features $|\mathcal{O}_n \cap \mathcal{F}_i|$. This step is repeated for the remaining feature subset $\mathcal{O}'_n = \mathcal{O}_n \backslash (\mathcal{O}_n \cap \mathcal{F}_i)$, which means that another feature set $\mathcal{F}_j$ is chosen according to $|\mathcal{O}'_n \cap \mathcal{F}_j|$ and so on. The decomposition terminates if the ratio $|\mathcal{O}'_n|/|\mathcal{O}_n| < \tau_d$ falls below the threshold τ_d (usually $\tau_d = 0.2$) or no further feature set is found. This top-down decomposition is repeated several times, and the final result $(\hat{\mathcal{G}}_n, \hat{\theta}_{\mathcal{G}_n})$ corresponds to the sample that maximizes $p(\mathcal{G}_n, \theta_{\mathcal{G}_n} | \mathcal{O}_n, \mathcal{G}^{n-1}, \theta_{\mathcal{G}}^{n-1})$. This learning by recognition leads to efficient hierarchies, where the recognition process significantly reduces the number of necessary iteration to find a good decomposition. Actually, in our experiments we found that about 10 iterations are typically sufficient.

ONLINE LEARNING($\mathcal{O}_n, \mathcal{G}^{n-1}, \theta_{\mathcal{G}}^{n-1}$)

1. Initialize the auxiliary feature set $\mathcal{A}$ and the set of state vectors z

    a. Initialize the auxiliary feature set $\mathcal{A}$ with the feature sets $\mathcal{F}$ provided by the set of hierarchies $\mathcal{G}^{n-1}, \theta_{\mathcal{G}}^{n-1}$
    b. Associate the feature set $\mathcal{O}_n$ to a level ℓ_n according to eq. 3.33.
    c. Add the feature set $\mathcal{O}_n$ to the auxiliary feature set $\mathcal{A}^{\ell_n}$.
    d. Add a randomly initiated state vector to $z^{(\tau)}$ (since the elements of $\mathcal{A}$ associated to $\mathcal{F}$ are already decomposed, just one state vector for $\mathcal{O}_n$ has to be added).

2. Use the proposal distribution $q(z^*|z^{(\tau)})$ to get a new state vector
3. TOP-DOWN DECOMPOSITION($z^*, \mathcal{A}^{\ell_1}, ..., \mathcal{A}^{\ell_{top}}$)
4. Accept the new state vector $z*$ and the posterior $p(\mathcal{G}_n, \theta_{\mathcal{G}_n}|\mathcal{O}_n, \mathcal{G}^{n-1}, \theta_{\mathcal{G}}^{n-1})$ with the acceptance rate given in eq. 4.15.
5. Repeat steps 1-4 (skip step 1d) for T_{on} iterations
6. Build the hierarchical structure using the sample that maximizes $p(\mathcal{G}, \theta_{\mathcal{G}}|\mathcal{O})$.

    a. Each element of $\mathcal{A}$ (not associated to $\mathcal{F}$) corresponds to a new node x_i that is added to $\mathcal{V}$
    b. The edges $e \in \mathcal{E}$ are defined between x_i and its children, which are given by the decomposition structure.
    c. Sharing links $e \in \mathcal{S}$ are established based on the shared structure and the determined representatives.

Algorithm 3 Online learning: Given a training instance $\mathcal{O}_n$ and a set of hierarchies $\mathcal{G}^{n-1}, \theta_{\mathcal{G}}^{n-1}$, online learning builds a new hierarchy and adapts it to the already learned set of hierarchies.

4.2.3 Parameter Learning

In the previous section, we describe how to learn the hierarchical structure $\mathcal{G}$; the corresponding set of parameters $\theta_{\mathcal{G}}$, which contains the parameters of the pairwise potential and observation functions, is yet undefined. We model the pairwise potential functions as simple Gaussian models with one relative position vector r_{ji}

$$\psi_j(x_i, x_j = \tilde{x}_j^{(\ell)}) = \mathcal{N}(x_i; (\tilde{u}_j^{(\ell)}, \tilde{v}_j^{(\ell)}) + r_{ji}, \Lambda_i) \qquad (4.19)$$

The relative position vector r_{ji} is determined using the auxiliary feature set $\mathcal{A}$, where r_{ji} between parent node x_j and its child node x_i is defined as

$$r_{ji} = center(\mathcal{A}_i) - center(\mathcal{A}_j) \qquad (4.20)$$

the function $center(\mathcal{A}_i) = 1/N_{\mathcal{A}_i} \sum_{j \in \mathcal{A}_i} \mu_j$ calculates the center of the features. More difficult is the appropriate estimation of the covariance Λ_i. Since statistical information is not available, the covariance has to be predefined. We choose a diagonal covariance $\Lambda_i = \sigma_{\ell_i} I$ with $\sigma_{\ell_i} = \alpha_\sigma(\beta_\sigma)^{\ell_i}$, typically $\alpha_\sigma \in \{0.5 - 1.5\}$ and $\beta_\sigma \in \{1.05 - 1.2\}$. This guarantees that features at lower levels are strongly connected, while at higher levels the parts are loosely connected. For shared nodes however, statistical information is available since multiple instances represent one object class. Assume we have a sharing group $\mathcal{S}$ with representative node r (see Sec. 3.7.3). For one specific decomposition we get therefore a set of relative position vectors $r_{j1}, ..., r_{j|\mathcal{S}|}$. We can calculate the covariance representing the displacement of the subparts by

$$\Lambda_r = \left(\sum_{i \in \mathcal{S}} \left(\Lambda_i + r_{ji} r_{ji}^T \right) \right) - r_{jr} r_{jr}^T \qquad (4.21)$$

The observation potentials are defined as described in Sec. 3.5.1.

4.2.4 Scale and Rotation Invariant Representation

For every object, view or articulation dependent instance we build a scale and rotation invariant representation. As already discussed in Sec. 3.7.1.1, the overall appearance of the instances stays the same during scaling and rotation. That is the reason why we apply structure sharing among all scales and rotations. As before, each scale and rotation is represented by one root node.

The structure sharing allows us to find an efficient decomposition once and use it for all other scale and rotation instances. The building scheme is similar to online learning. While during online learning different hierarchical decompositions are tested and the one that maximizes the reusability is chosen, we are using exactly the same hierarchical decomposition. Just the potential functions associated to the edges have to be adapted (see Sec. 3.6):

- **Scale:** The relative position vectors r_{ji} have to be scaled $r_{ji}^s = s r_{ji}$.
- **Rotation:** The relative position vectors r_{ji} have to be rotated $r_{ji}^\alpha = R^\alpha r_{ji}$.

where s denotes the scaling factor and R^α a rotation matrix. The learning proceeds top-down and compares the rotated and scaled feature sets to those already included in the set of hierarchies. If a new node x_j and its associated feature set $\mathcal{F}_j$ fulfills the sharing criterion with an already learned node x_i, an edge l_{ij} is added to $\mathcal{L}$. The top-down processing terminates when all new nodes are linked to already learned nodes. During scaling, the number of features has also to be adapted $N_i^s = s N_i$, so that the associated level might change according to eq. 3.33.

Another important property, which has to be regarded during learning, is that the similarity between the instances increases as the objects are scaled and downsized. Referring to the sharing criterion this increasing similarity allows to enlarge the number of shared primitives, so that the overall number of primitives decreases.

4.2.5 Learning of the Similarity Hierarchy

There are two reasonable approaches to construct the similarity hierarchies based on already learned compositional hierarchies:

1. One common solution to learn a similarity hierarchy is by means of a hierarchical clustering algorithm [169, 76, 75]. The clustering starts at the bottom (of the finest compositional hierarchy) and clusters iteratively the samples (here nodes and their associated feature sets) proceeding bottom-up. Each formed cluster defines a node of the next higher level. The k-means algorithm is one common clustering approach, but unfortunately it needs the number of clusters to be known. Furthermore, the algorithm needs feature vectors and an appropriate metric, typically Euclidean, as input in order to calculate distances and recalculate the position of the centroid. However, since in our case the input are feature sets, the formulation as a feature vector of a fixed size demands other transformations. Thus, clustering approaches based on distance matrices like the density-based clustering algorithm [52] are more appropriate. The input matrix can be calculated using the distance function $dist(\mathcal{A}_i, \mathcal{A}_j)$ as introduced in Sec. 3.5.1.

2. The scale space representation has another attractive property that facilities the learning of similarity hierarchies as already described in Sec. 3.7.4. Since our hierarchical model contains already scaled versions of each object and its parts, we can simply use the same compositional hierarchy and shift it horizontally according to the level of detail. The nodes of the compositional level ℓ and level of detail ℓ_d thus corresponds to nodes of the compositional level $\ell + 1$ and level of detail $\ell_d + 1$. The similarity edges between the different levels of detail are established according to the similarity criterion. Due to the scale space representation the feature set at the finer level has to be scaled before calculating the similarity.

Chapter 5
Object Recognition

In this chapter, we will apply the hierarchical framework proposed in Chap. 3 and the corresponding learning proposed in Chap. 4 to object recognition and present experimental results (example shown in Fig. 5.1). The results are based on the Columbia Object Image Library (COIL-100) [165], which provides color images of 100 objects each captured from 72 views.

5.1 Related Work

We will now give an overview of related work. Since the related work concerning hierarchical models is already reviewed in Sec. 3.2, we will now focus on the low-level feature extraction techniques from the literature. One class of feature extraction tries to find points in the image, which are unique within their neighborhood. Moravec [152] detected *points of interest*, which are defined as points where locally large intensity variations are present in every of the four main directions. The operator calculates the similarity between each point and its neighborhood, where the similarity is measured by taking the sum of squared differences between the patch centered on the current pixel and patches centered on neighbors. Harris and Stephens [86] proposed a corner detector that uses an auto-correlation matrix. The analysis of the eigenvalues of the matrix allows to distinguish between corners, edges and uniform regions. Similar approaches were applied to multiple scales by e.g. Lindeberg [133] for the detection of blobs, corners, edges and ridges.

Another common feature class are edges, which have discontinuities in at least one direction. The well-known Sobel operator calculates approximations of the derivatives in horizontal and vertical direction using convolution kernels. These kernels additionally smooth the image perpendicular to the derivative direction in order to reduce the influence of noise. They can be seen as simple edge templates which are compared with the image [240]. As the image noise increases the template size has to increase as well. This also means a larger variability of possible edges within the operator window. Hueckel [97] proposed to use a set of orthogonal basis functions to represent the variability of possible edges. Since the edge detection used a

© Springer International Publishing Switzerland 2015
J. Spehr, *On Hierarchical Models for Visual Recognition & Learning of Objects, Scenes, & Activities*,
Studies in Systems, Decision and Control 11, DOI: 10.1007/978-3-319-11325-8_5

computational expensive optimization several improvements were proposed. Mero and Vassy [146] showed that the direction of the edge can be efficiently calculated based on two simple basis functions. Burow and Wahl [28, 240] extended this approach and proposed an efficient calculation of the edge position.

The *Canny edge detector* is another widely used edge detector [29]. It was designed for

- Good detection. The detector should achieve a high true positive rate and the probability of correctly marking real edge points should therefore be high.
- Good localization. The detected edge points should be as close as possible to the real edges.
- Only one response to a single edge. The detector should achieve a high true negative rate and the probability of falsely marking nonedge points should therefore be low.

A simplified version of the detection procedure can be divided into the following four steps:

1. Convolve the input image with a Gaussian kernel in order to smooth the signal.
2. Compute the gradients. The first partial derivatives can e.g. be calculated using the Sobel masks.
3. Perform non-maximum suppression. Depending on the edge direction, gradients which are not local peaks are set to zero.
4. Hysteresis Thresholding. Two thresholds τ_{low} and τ_{high} are used to decide if a point is an edge feature or not. If the magnitude of a point lies below τ_{low}, it is rejected, if it is above τ_{high}, it is accepted. For a magnitude between τ_{low} and τ_{high} the corresponding point is accepted if it is connected to a neighbor with a magnitude above τ_{low}.

Another class of feature detectors is based on wavelets, which are related to the perception in the human visual system. Daugman [37] presented evidence that the 2d receptive-field profiles of simple cells in the mammalian visual cortex are well described by members of 2d Gabor wavelets, which are Gaussian kernel functions modulated by sinusoidal plane waves. For edge detection, typically a filter bank of Gabor wavelets with different frequencies and orientations is used [66].

5.2 Hierarchical Representation

5.2.1 Low-Level Image Features

In our framework, the aim of the feature extraction is a dense representation at the lowest level. Instead of using a sparse representation, gathered for example by a corner detector, and using descriptors with a high distinctiveness, we aim to use a dense representation and a less distinctive descriptor since this guarantees that the reusability of parts is high. Descriptors like SIFT are having a high distinctiveness and are thus in general not reusable across classes. Without limiting the generality

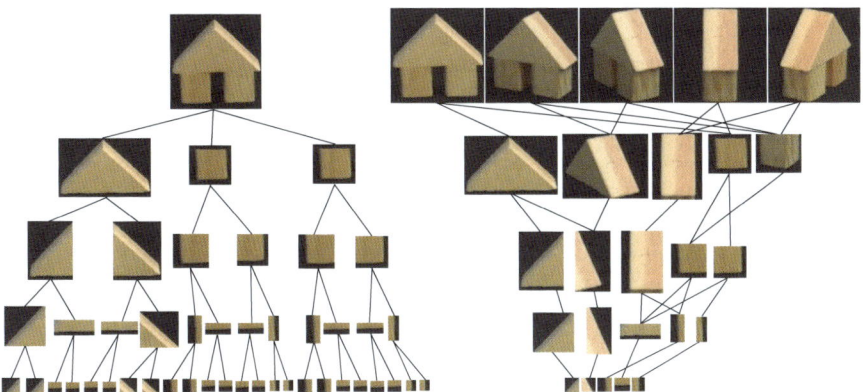

Fig. 5.1 Example of the hierarchical decomposition of an object. The object is composed of parts, and parts are, in turn, composed of visual primitives (left). Instead of evaluating all poses of the object separately we take advantage of the circumstance that parts share visual primitives (right).

of our approach, we will, in this monograph, use the Canny edge detector with non-maximum suppression. The recognition as well as the learning framework, however, is suitable for other feature classes. We choose the Canny detector due to its sparseness (which is simultaneously a dense edge representation), robustness, simple and fast calculation. The Canny edge detector, as described in Sec. 5.1, delivers a set of feature positions $\mu_1, ..., \mu_n$. As mentioned in Sec. 3.5.1, we will use features in our hierarchy as local image evidence; we are more interested in good locality performance accepting a worse distinctiveness of the descriptor. In order to guarantee locality, we use a simple color descriptor in the direct neighborhood of the edge. The color values are calculated as the average color value on each side of the edge. We distinguish between two types of edges. On the one hand we have edges that are element of the object's silhouette and thus represent the object border. In this case we are just interested in the color information on the object side. On the other hand we have edges due to color changes within the object silhouette, where we are interested in both color values. The low-level feature descriptor is defined as $\delta = (\delta x, \delta y, c_1, c_2)$, where c_1, c_2 are the color values and $(\delta x, \delta y)$ is the gradient in x- and y- direction. We employ the L*u*v* color space since it is designed to optimally approximate a perceptually uniform color space and has linear mapping properties [32]. L* is the lightness coordinate, u* and v* are the chromaticity coordinates. Thus, our low-level representation of the image is given by the feature set $\mathcal{F} = \{f_1, ..., f_{N_f}\}$, where each feature is defined as a pair $f = (\mu, \delta)$ with the position $\mu \in \mathbb{R}^2$, and the descriptor δ.

We define the distance between two features as

$$d_{w_p, w_c, w_a}(f_1, f_2) = \left((w_p \Delta_p(\mu_1, \mu_2))^2 + (w_c \Delta_c(\delta_1, \delta_2))^2 + (w_a \Delta_a(\delta_1, \delta_2))^2\right)^{0.5}$$

$$(5.1)$$

with $\Delta_p(\mu_1, \mu_2) = \|(x_{f_1} - x_{f_2}, y_{f_1} - y_{f_2})\|$, $\Delta_c(\delta_1, \delta_2) = \|(L_{f_1} - L_{f_2}, u_{f_1} - u_{f_2}, v_{f_1} - v_{f_2})\|$ and $\Delta_a(\delta_1, \delta_2) = \|(\delta x_{f_1} - \delta x_{f_2}, \delta y_{f_1} - \delta y_{f_2})\|$, where $w_p, w_c, w_a \in [0, 1]$ are weights, which will be discussed later. As we have seen, the distance between two feature sets will be needed during the learning process. We define the distance between two feature sets $\mathcal{F}_1 = \{f_1^1, ... f_{N_{\mathcal{F}_1}}^1\}$ and $\mathcal{F}_2 = \{f_1^2, ..., f_{N_{\mathcal{F}_2}}^2\}$ as an extension of the modified Hausdorff distance as

$$dist(\mathcal{F}_1, \mathcal{F}_2) = \max(h_{mhd}(\mathcal{F}_1, \mathcal{F}_2), h_{mhd}(\mathcal{F}_2, \mathcal{F}_1)) + \zeta \frac{|N_{\mathcal{F}_1} - N_{\mathcal{F}_2}|}{N_{\mathcal{F}_1} + N_{\mathcal{F}_2}} \quad (5.2)$$

with

$$h_{mhd}(\mathcal{F}_1, \mathcal{F}_2) = \frac{1}{N_{\mathcal{F}_1}} \sum_{f_i \in \mathcal{F}_1} \min_{f_j \in \mathcal{F}_2} d_{w_p, w_c, w_a}(f_i, f_j) \quad (5.3)$$

where the parameter ζ additionally penalizes sets that differ substantially in the feature number. Please note, that in the following we will assume that two feature sets are aligned when we calculate $h_{ext}(\mathcal{F}_1^{align}, \mathcal{F}_2^{align})$. For this we align the feature sets according to their center $\mathcal{F}^{align} = \{\mu_i - center(\mathcal{F}), \delta_i\}_{i=1,...,N_{\mathcal{F}}}$. Of course other techniques, like e.g. RANSAC, would also be appropriate and would cope better with outliers, but are not further investigated in this monograph.

5.2.2 Unsupervised Learning

In this section we describe, how to learn an optimal representation of a set of n object view instances $\mathcal{O} = \{\mathcal{O}_1, ..., \mathcal{O}_n\}$ which maximizes the reusability of parts. We apply the offline learning, as described in Sec. 4.2.1, and use the Metropolis sampler in a top-down manner to get samples from the posterior distribution $p(\mathcal{G}, \theta_{\mathcal{G}}|\mathcal{O})$. Each sample represents one set of hierarchies, while the final learning results correspond to the sample which maximizes $p(\mathcal{G}, \theta_{\mathcal{G}}|\mathcal{O})$.

Segmentation: In order to guide the decomposition and restrict the set of possible decompositions to reasonable ones, a set of state vectors $z = \{z_\ell\}_{\ell=1}^{\ell_{top}}$ is used. A state vector $z_i = (w_i, \mathcal{K}_i)$ is associated to each element $\mathcal{A}_i$ of the auxiliary feature set. It includes a weighting vector $w_i = (w_p, w_c, w_a)$, $w_p, w_c, w_a \in [0, 1]$, and a set of mean features $\mathcal{K}_i = (s_1, ..., s_k)$. The weighting vector scales the position, the angular and the color components of the feature and thus allows to emphasize different feature properties. During the segmentation each feature $f_j \in A_i$ is assigned to the nearest mean feature $\hat{k}_j = \operatorname{argmin}_{k=1,...,n} d_{w_p, w_c, w_a}(f_j, s_k)$. The assignment allows a segmentation of the feature set based on the state vector z. The mean features are randomly chosen from the feature set A_i, while the number of mean features is uniformly distributed within the interval $\{2, ..., k_{max}\}$. Some examples of the segmentation results are shown in Fig. 5.2. In Fig. 5.2(c), the feature set is segmented into $k = 2$ and $k = 3$ parts according to a weighting vector $w_i = (0, 1, 0)$, which emphasize the color components. A spatial segmentation can be seen in Fig. 5.2(b), where $w_i = (1, 0, 0)$ weights the spatial component high. And in Fig. 5.2(d) one can see segmentation results according to the gradient, thus $w_i = (0, 0, 1)$.

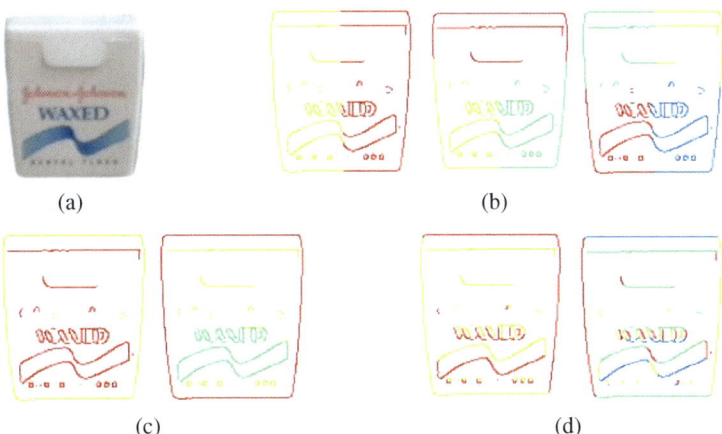

Fig. 5.2 Segmentation results according to ((a) original image): (b) space, (c) color, and (d) gradients (see text for a detailed description).

The set of state vectors defines the whole forest of trees, so that the set of hierarchies depends on z

$$(\mathcal{G}, \theta_{\mathcal{G}}) = (\mathcal{G}(z), \theta_{\mathcal{G}}(z)) \tag{5.4}$$

Proposal Function. We use a symmetric proposal function $q(z^*|z^{(\tau)}) = q(z^{(\tau)}|z^*)$, so that the sampler becomes a Metropolis sampler with an acceptance rate for a new candidate according to eq. 4.15. For the weighting part w_i of the state vector, a Gaussian distribution is used

$$w^* \sim \mathcal{N}(w^{(\tau)}; \sigma_w^2 I) \tag{5.5}$$

The proposal distribution of the set of mean features $\mathcal{K}_i = (s_1, ..., s_k)$ is divided into two parts. First, the number of mean features is updated

$$k^* \sim k^{(\tau)} + \mathcal{U}(-1, 1) \tag{5.6}$$

where we limit the number to the interval $[2, k_{max}]$. If the proposal distribution decreases the number of mean features, a random feature is chosen and removed from $\mathcal{K}^\tau$. On the other hand, if the number increases we randomly choose a new feature from A_i and add it to $\mathcal{K}^\tau$. Here, we avoid that one feature is chosen twice. All other features contained in $\mathcal{K}^\tau$ are modified using a Gaussian distribution

$$s^* \sim \mathcal{N}(s^{(\tau)}; \sigma_s^2 I) \tag{5.7}$$

The sampling proceeds in a downward decomposition scheme, starting at the highest level of the auxiliary feature set. Initially, this feature set contains just the root nodes of the training instances. As described in Sec. 4.2.1, samples are generated

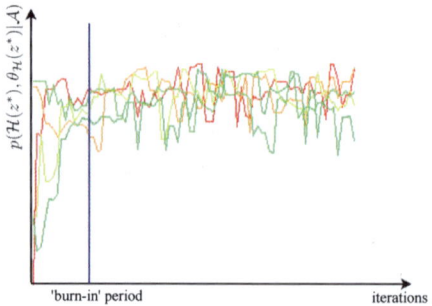

Fig. 5.3 Example curves of the posterior probability during the sampling process. As can be seen, after the burn-in period the sampler generates likely samples.

for each level separately, while the samples always depend on the results of the next higher levels. During the sampling of the set of state vectors z_{ℓ_i} at level ℓ_i, new samples are generated in several iterations. Usually, in each iteration all state vectors are modified using the proposal functions. However, this leads to an inefficient sampling scheme since if the new sample will not be accepted, all intermediate calculations will be thrown away. Unfortunately, even in this case, it is very likely that among the rejected intermediate results valid state vectors remain. We therefore apply the sampling in each iteration on a subset of z_{ℓ_i}. The subset is randomly chosen and contains 30% of the state vectors of z_{ℓ_i}. Example curves can be seen in Fig. 5.3. In general, we use a burn-in period of 20 samples and omit all samples generated in this phase.

5.3   Results

We use the coil-100 database to experimentally evaluate our proposed hierarchical framework and show the benefits. The evaluation is divided into a learning and a detection section. In the first one, we will investigate the different learning approaches proposed in Sec. 4 (offline and online learning). In the detection section, we investigate the different message passing schemes and show how our approach overcomes limitations of previous approaches concerning occlusions or clutter.

5.3.1   Learning

We evaluate the learning approach based on different objects from the coil database. The first object we choose is a cup (COIL object id 25). This object is especially interesting since it is rotationally symmetric and therefore view point independent. In this case, we expected the degree of sharing to be high. The second object we choose is a toy block tower, where the view instances are less similar but where both sides look the same (COIL object id 51). The third object is a package, that has a simple shape but complicated colored labels (COIL object id 65). All objects are learned using the same framework with exactly the same learning parameters.

5.3.1.1 Offline Learning

We apply the offline learning approach to objects of the COIL database. In order to demonstrate the influence of different sharing degrees we investigate several learning runs while varying the sharing properties. The sharing threshold τ_e is the main component that is used to decide if two parts can be shared or not. We therefore investigate different values for τ_e. Although the modified Hausdorff distance is already normalized with respect to the number of particles and we could thus apply exactly one threshold to all levels, we found, that larger thresholds at higher levels lead to more efficient hierarchies. Therefore, we enlarge the threshold at higher levels in order to improve the sharing of object parts. The threshold τ_e depends on the level according to

$$\tau_e(\ell) = \alpha_{\tau_e}(\beta_{\tau_e})^\ell \tag{5.8}$$

with $\alpha_{\tau_e} = 0.1$ and $\beta_{\tau_e} = \in \{1.1 - 1.6\}$.

In Tab. 5.1 we summarized the learning results for the rotationally symmetric object 25. For different sharing parameters β_{τ_e} the average self-difference between the learned representation $\mathcal{F}_i$ and the observation $\mathcal{O}_i$ is shown. Ideally, this difference equals zero. However, the difference increases as the parameter β_{τ_e} and thus sharing enlarges. The average difference between different instances is also shown. As can be seen, the difference has also a relatively low value owing to the fact that the object is rotationally symmetric and all views are similar. This also becomes apparent in Fig. 5.9, where the similarity matrices are shown. The hierarchical structure on the other hand gets more and more simplified as the threshold increases. While for $\beta_{\tau_e} = 1.1$, 1193 nodes and 4619 edges are needed, the number is reduced to approximately one tenth for $\beta_{\tau_e} = 1.6$. Since each edge requires the calculation of at least two messages (upwards and downwards) the number of edges crucially determines the computational effort. We use a reference hierarchy that connects each root node directly with all of its low-level features. The number of edges $|\mathcal{E}_{\mathcal{G}_{ref}}|$ corresponds in this case to the number of low-level features. We define the sharing degree as the ratio $|\mathcal{E}_{\mathcal{G}_{ref}}|/|\mathcal{E}_{\hat{\mathcal{G}}}|$. A sharing degree of 2 thus means, that the number of messages is halved. For object 25 the sharing degree varies between 14 and 121. Another interesting parameter is the average number of parent nodes, which corresponds to the ratio $|\mathcal{E}_{\hat{\mathcal{G}}}|/|\mathcal{V}_{\hat{\mathcal{G}}}|$. Thus, this measure also considers the number of nodes, while the sharing degree just considers the number of edges. If the average number of parent nodes is high, the automatically learned nodes are efficiently reused by multiple parents. This means in general, that these learned nodes are good guesses. In the worst case, where a node has just one parent, the likelihood for a good guess is low and these nodes are in general very object specific. The log posterior reaches its maximum for $\beta_{\tau_e} = 1.3$, where we used the parameters $\gamma_1 = 0.1$, $\gamma_2 = 0.1$, $\gamma_d = 20.0$ to calculate the log posterior according to eq. 4.11 and 4.12.

The corresponding hierarchy can be seen in Fig. 5.5 (not all primitives are shown). The root nodes are highlighted by a green bounding box, as can be seen they are distributed at level 19 and 20. The fact that the root nodes do not have to lie on the same level is an important property of our hierarchical representation. The nodes that are shared among different configurations are marked by a black circle. Especially at the low levels, the number of shared primitives is high. It is interesting that the result generated by the MAP sampler segments the cup according to its shape on the one hand and its edges on the other hand (level 17 and 18).

The results for object 51 are shown in Tab. 5.2. The average difference between feature sets of different views is much higher than for object 25. The same applies to the number of nodes and edges, which are approximately twice the number for object 25. The similarity matrices are given in Fig. 5.10 and show that the opposite sides of the object are equal. Starting at a value for β_{T_e} of about 1.3, the matrices show more and more artifacts. As expected, the sharing degree is much lower than for the rotationally symmetric object. It ranges between 4 and 130 (except for $\beta_{T_e} = 1.6$). The average number of parents is similar to object 25. The best hierarchy ($\beta_{T_e} = 1.4$) according to the log posterior measure is shown in Fig. 5.6. The feature sets associated to the root nodes consist of less features than for object 25, thus they are distributed at level 18 and 19.

Tab. 5.3 gives the results for object 65. The difference between feature sets of different views is very high due to the large number of features and their large variations. The corresponding similarity matrices are shown in Fig. 5.11. The degree of sharing as well as the average number of parent nodes is similar to object 51. The best set ($\beta_{T_e} = 1.3$) of hierarchies is shown in Fig. 5.7. The root nodes are distributed on level 17 to level 21. The simple back-side of the box (without texture) is assigned to level 17 and the complex front including the labels (with texture) is assigned to level 21.

We also apply the offline learning to all object categories simultaneously. For each object we chose the front view and trained the hierarchies with different sharing parameters β_{T_e}. Tab. 5.4 gives the results for object 1-100. The average difference between the objects is much larger than in the previous experiments. The similarity matrices in Fig. 5.12 are clearly highlighting the diagonal elements. However, for $\beta_{T_e} > 1.3$ the confusion increases significantly. Surprisingly, the degree of sharing reaches similar values as for object 51 and 65. However, mainly simple primitives (level < 15) are shared and most of the complex high-level parts are object specific. The final set of hierarchies is shown in Fig. 5.8 for $\beta_{T_e} = 1.4$, where the log posterior reaches its maximum.

The distribution of the root nodes is also summarized in Fig. 5.13. Object 31 is the most complex object with 2462 features, object 94 is the most simple object with 296 features. As can be seen, the root nodes are mainly distributed on level 19 and 20. The overall distribution of primitives is shown in Fig. 5.14. Each diagram shows the level distribution for different sharing parameters and for one specific training example. For $\beta_{T_e} = 1.1$, the curves show especially high values at level 5-8. For $\beta_{T_e} = 1.2$ these large numbers are reduced, but interestingly the number of parts is slightly increased at higher levels (9-12). The reason for this is, that the reduced set of simple primitives (level 5-8) causes larger sets of more complex features (level 9-12). Among these features are three groups of primitives.

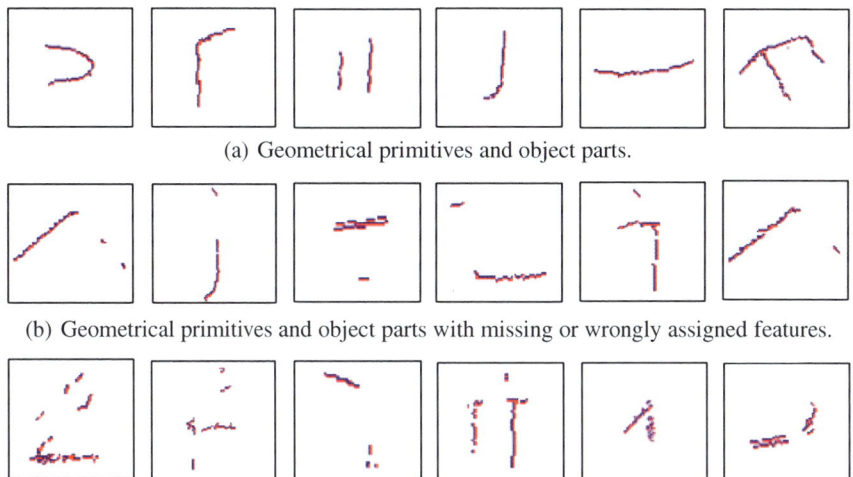

(a) Geometrical primitives and object parts.

(b) Geometrical primitives and object parts with missing or wrongly assigned features.

(c) Invalid primitives due to unreasonable segmentations and noisy low-level features.

Fig. 5.4 Groups of primitives.

- The first one consists of geometrical primitives like corners, edges, parallel lines, and of more complex features, which correspond to parts of objects (see Fig. 5.4(a)). While the geometrical primitives have a high reusability, the object parts are in general very object specific and thus have a low reusability. In our framework the different degrees of sharing are regarded by the importance weights introduced in Sec. 3.8.2. Since this first group represents valid features we would ideally just want features of this group.
- The second one represents features which are actually as good as the ones of the first group (see Fig. 5.4(b)). Unfortunately, due to missing or wrongly assigned features their difference to other primitives is too large. As a consequence, they fall not in one cluster with other primitives and are therefore not shared. This makes this kind of primitive very costly and the aim should be to reduce the number of primitives of this group by assigning them to the elements of the first group. One solution might be the use of the RANSAC approach during the alignment. We expect that this improves the calculation of the distance function as the RANSAC approach was especially designed to deal with outliers.
- The last group are invalid primitives, which should actually be avoided by means of the state vectors z guiding the segmentation (see Fig. 5.4(c)). They result from unreasonable segmentations and noisy low-level features, and are in general object and view specific, i.e. they have just one parent node. In our current implementation, which is used for evaluation purposes, we are not detecting and

blocking the elements of this group explicitly, although this might be reasonable. Since the elements of the third group are often characterized by a small number of features which are spatially widely dispersed, one could for example use heuristics which measure the ratio of the spatial expansion to the number of primitives.

Examples of shared features are shown in Fig. 5.15 - 5.18 . Fig. 5.15 shows some clusters and their elements for object 25. As previously mentioned, the degree of sharing is very high. The views share the same silhouette and the same circular primitives. The representative node for each cluster is determined according to eq. 3.37 and highlighted by a black box. For each element of the cluster (in red) we also aligned the representative (in blue). Fig. 5.16 and 5.17 show simple primitives like edges and corner but also the silhouette of the objects (Fig. 5.16(f) and Fig. 5.17(f)). The clusters of similar parts for object 1-100 in Fig. 5.18 show that mainly simple primitives are shared. If more complex parts are shared, as in Fig. 5.18(j), the number of parent nodes does not exceed two.

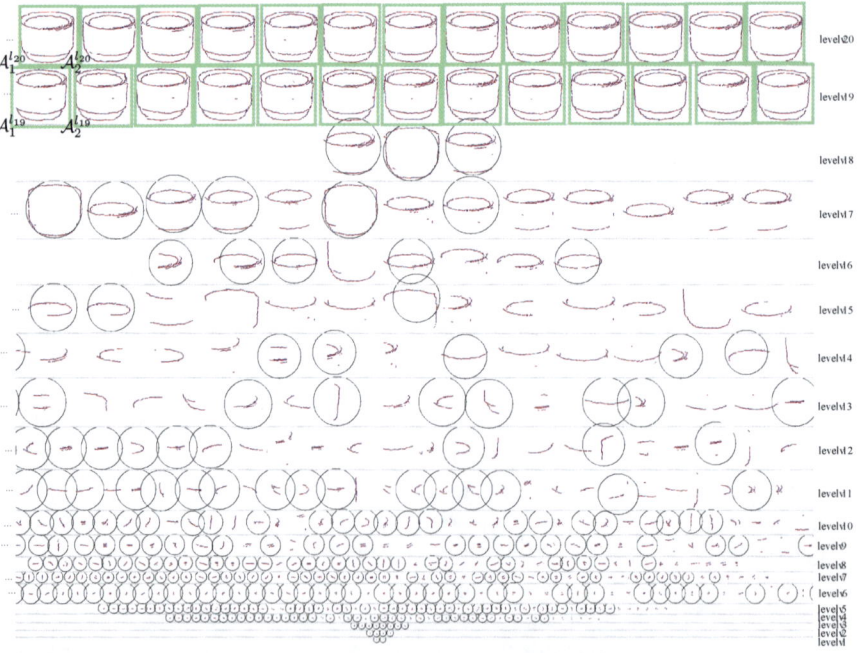

Fig. 5.5 Final learned set of hierarchies for the representation of object 25 (root nodes are highlighted by a green bounding box). Not all nodes are shown.

Table 5.1 Learning results for different sharing parameters β_{τ_e} for object 25 of the COIL database.

β_{τ_e}	avg. self distance	avg. distance	num. nodes	num. edges	sharing degree	avg. num. shared nodes	log posterior
1.1	1.14	3.55	1193	4619	13.98	3.87	-604.02
1.2	1.53	3.81	758	3332	19.38	4.40	-439.58
1.3	3.58	4.96	385	1569	41.15	4.08	-266.90
1.4	9.24	11.41	266	984	65.61	3.70	-309.80
1.5	11.86	12.70	189	698	92.49	3.70	-362.04
1.6	37.59	36.81	144	534	120.90	3.71	-819.57

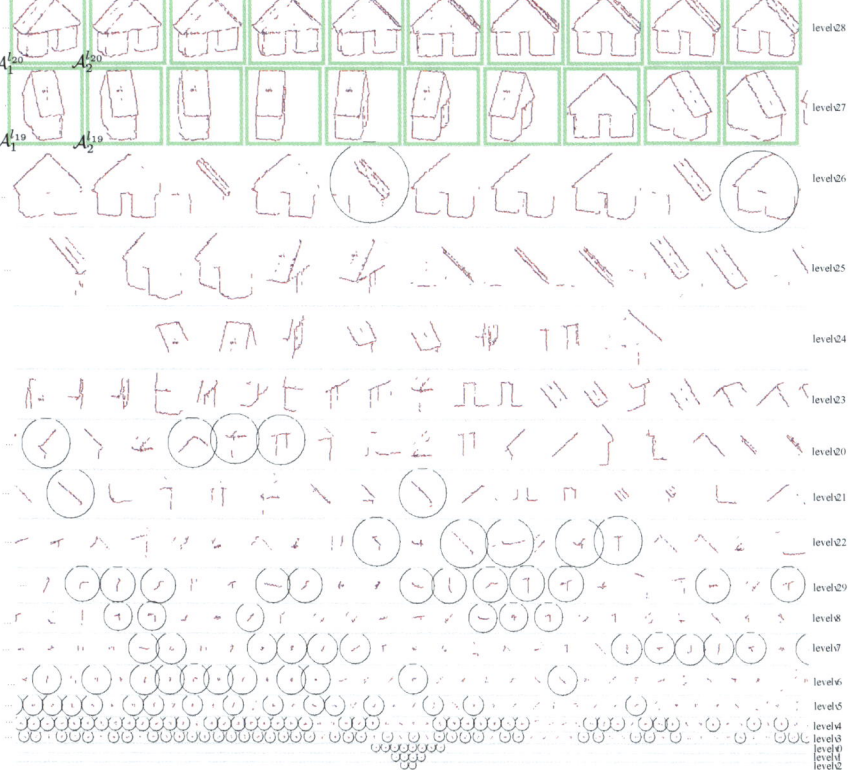

Fig. 5.6 Final learned set of hierarchies for the representation of object 51 (root nodes are highlighted by a green bounding box). Not all nodes are shown.

Table 5.2 Learning results for different sharing parameters β_{τ_e} for object 51 of the COIL database

β_{τ_e}	avg. self distance	avg. distance	num. nodes	num. edges	sharing degree	avg. num. shared nodes	log posterior
1.1	1.50	109.72	3073	11370	4.25	3.70	-1474.21
1.2	2.12	109.02	1845	5990	8.06	3.25	-825.84
1.3	4.23	107.02	855	3716	12.99	4.35	-541.62
1.4	11.74	104.31	437	1500	32.18	3.43	-428.47
1.5	30.04	110.07	219	721	66.96	3.29	-694.80
1.6	89.07	137.89	131	371	130.11	2.83	-1831.62

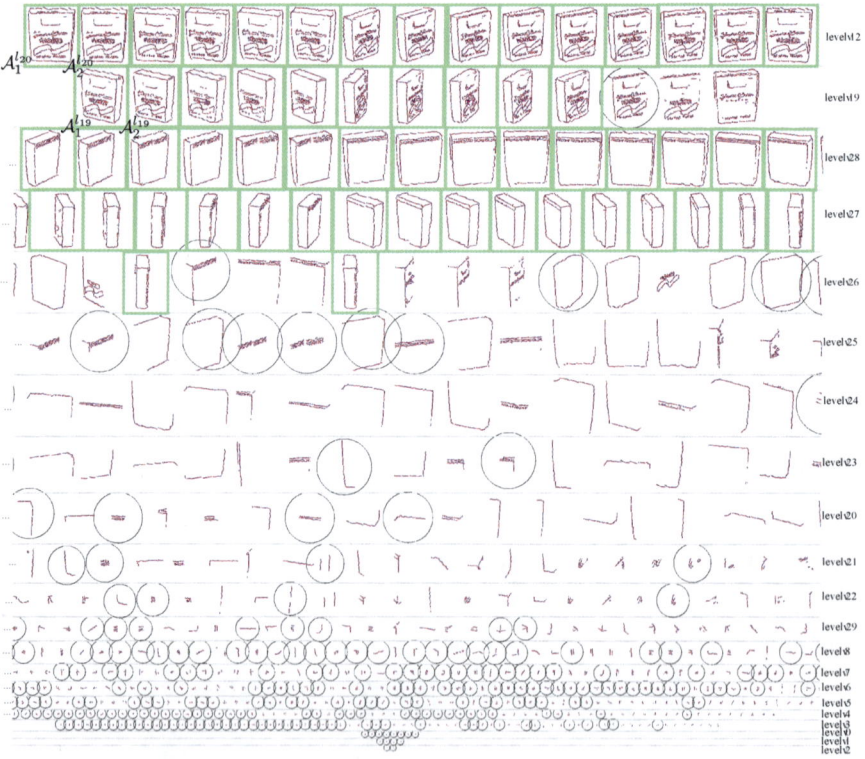

Fig. 5.7 Final learned set of hierarchies for the representation of object 65 (root nodes are highlighted by a green bounding box). Not all nodes are shown

Table 5.3 Learning results for different sharing parameters β_{τ_e} for object 65 of the COIL database

β_{τ_e}	avg. self distance	avg. distance	num. nodes	num. edges	sharing degree	avg. num. shared nodes	log posterior
1.1	0.42	198.07	3921	17037	3.82	4.35	-2104.20
1.2	1.39	197.01	1841	7693	8.46	4.18	-981.30
1.3	5.70	192.79	833	3265	19.93	3.92	-523.87
1.4	19.24	194.62	349	1320	49.30	3.78	-551.80
1.5	35.68	183.83	194	746	87.23	3.85	-807.67
1.6	97.19	173.13	131	447	145.57	3.41	-2001.56

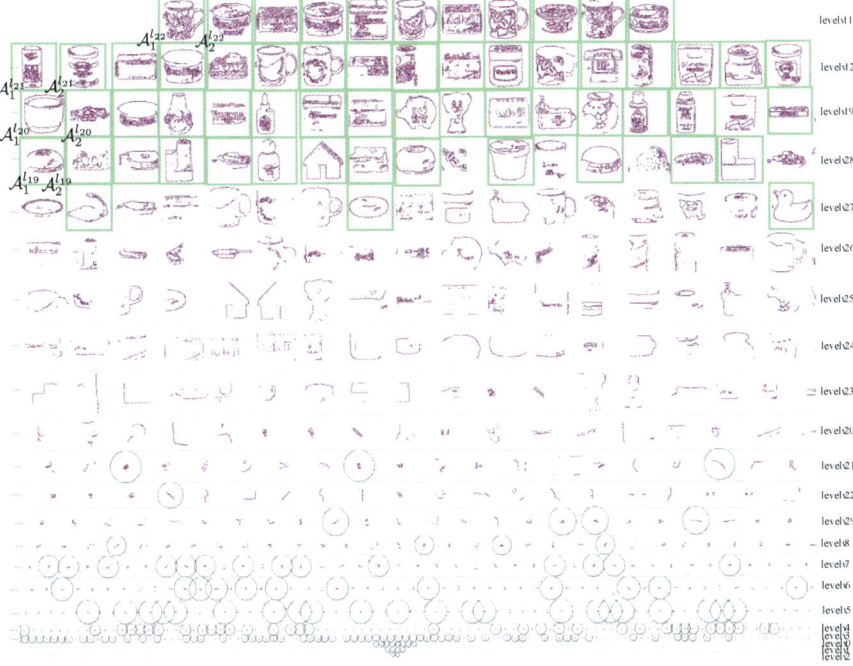

Fig. 5.8 Final learned set of hierarchies for the representation of objects 1 to 100 (root nodes are highlighted by a green bounding box). Not all nodes are shown.

Table 5.4 Learning results for different sharing parameters β_{τ_e} for object 1 to 100 of the COIL database

β_{τ_e}	avg. self distance	avg. distance	num. nodes	num. edges	sharing degree	avg. num. shared nodes	log posterior
1.1	0.81	220.77	8658	32882	3.55	3.80	-4170.29
1.2	0.99	220.03	5669	24304	4.81	4.29	-3017.19
1.3	3.65	216.67	2392	8687	13.44	3.63	-1180.99
1.4	16.22	204.93	885	3046	38.34	3.44	-717.58
1.5	36.60	196.73	473	1817	64.27	3.84	-961.01
1.6	58.00	156.14	261	960	121.65	3.68	-1282.05

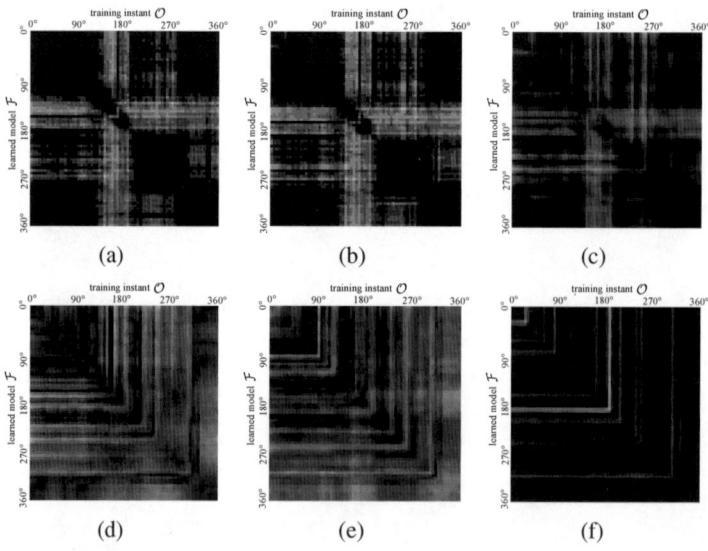

Fig. 5.9 Similarity matrices for object 25: (a) β_{τ_e}=1.1, (b) β_{τ_e}=1.2 ,(c) β_{τ_e}=1.3, (d) β_{τ_e}=1.4, (e) β_{τ_e}=1.5, and (f) β_{τ_e}=1.6.

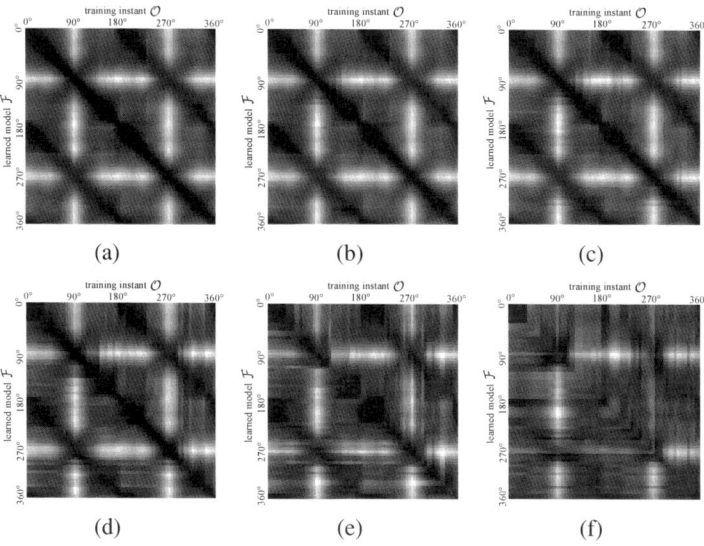

Fig. 5.10 Similarity matrices for object 51: (a) β_{τ_e}=1.1, (b) β_{τ_e}=1.2, (c) β_{τ_e}=1.3, (d) β_{τ_e}=1.4, (e) β_{τ_e}=1.5, and (f) β_{τ_e}=1.6.

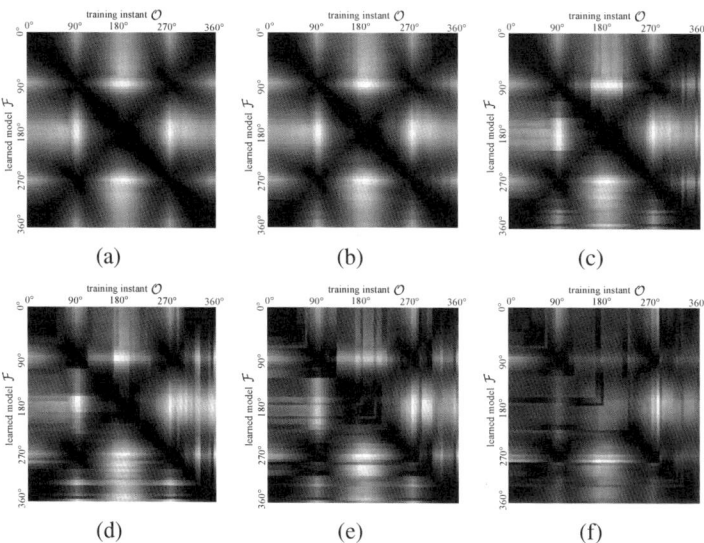

Fig. 5.11 Similarity matrices for object 65: (a) β_{τ_e}=1.1, (b) β_{τ_e}=1.2, (c) β_{τ_e}=1.3, (d) β_{τ_e}=1.4, (e) β_{τ_e}=1.5, and (f) β_{τ_e}=1.6.

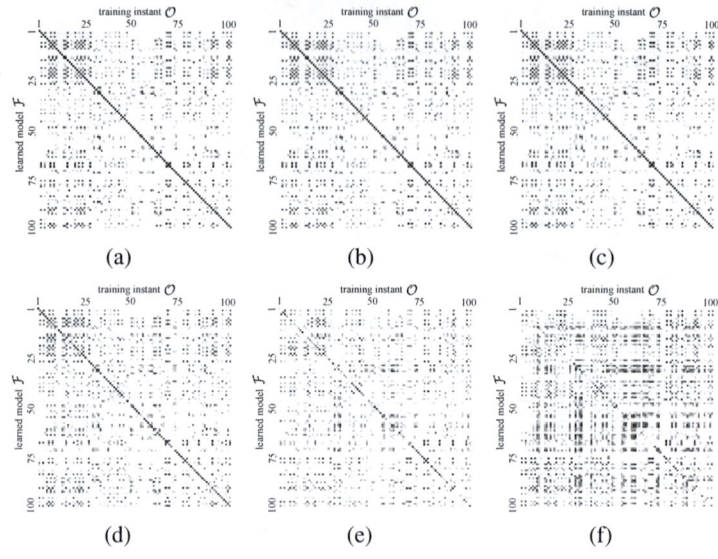

Fig. 5.12 Similarity matrices for object 1 - 100: (a) β_{τ_e}=1.1, (b) β_{τ_e}=1.2, (c) β_{τ_e}=1.3, (d) β_{τ_e}=1.4, (e) β_{τ_e}=1.5, and (f) β_{τ_e}=1.6.

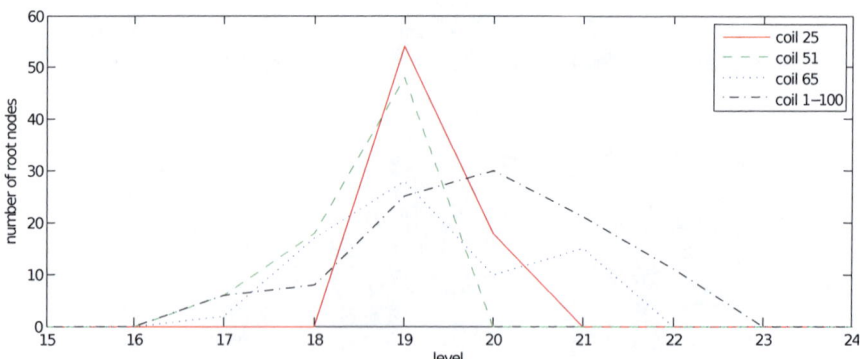

Fig. 5.13 Distribution of the root nodes depending on the hierarchy level

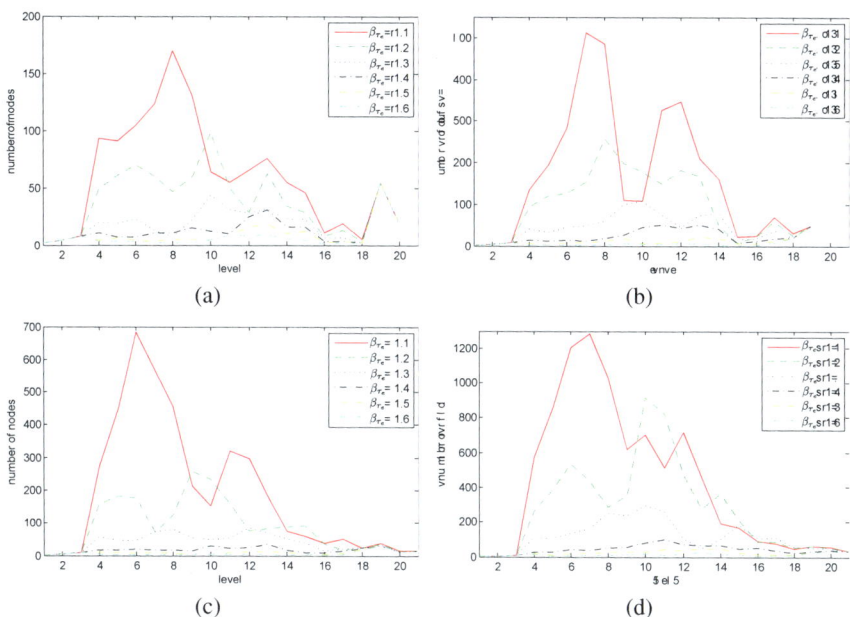

Fig. 5.14 Number of primitives as a function of the hierarchy level: (a) object 25, (b) object 51, (c) object 65, and (d) object 1-100.

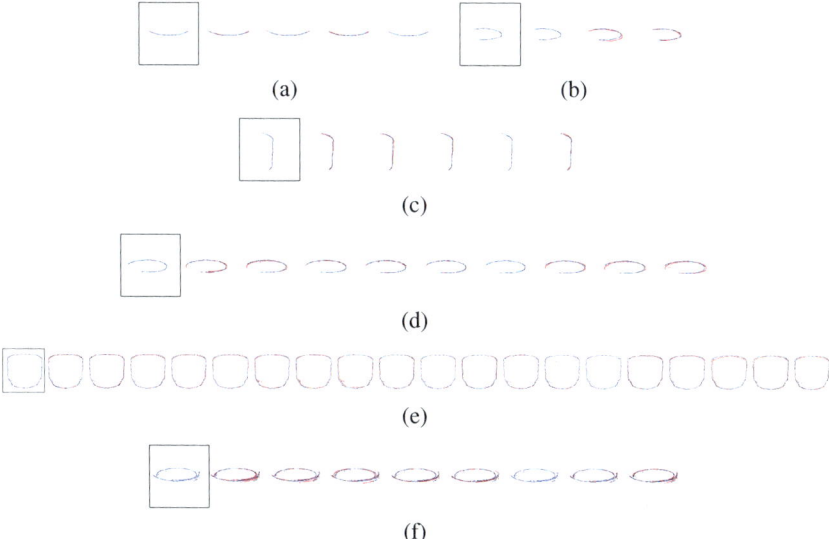

Fig. 5.15 Clusters of similar parts for object 25 (representative on the left in blue, cluster elements in red): (a) level 12, (b) level 13. (c) level 13. (d) level 14. (e) level 16. (f) level 16.

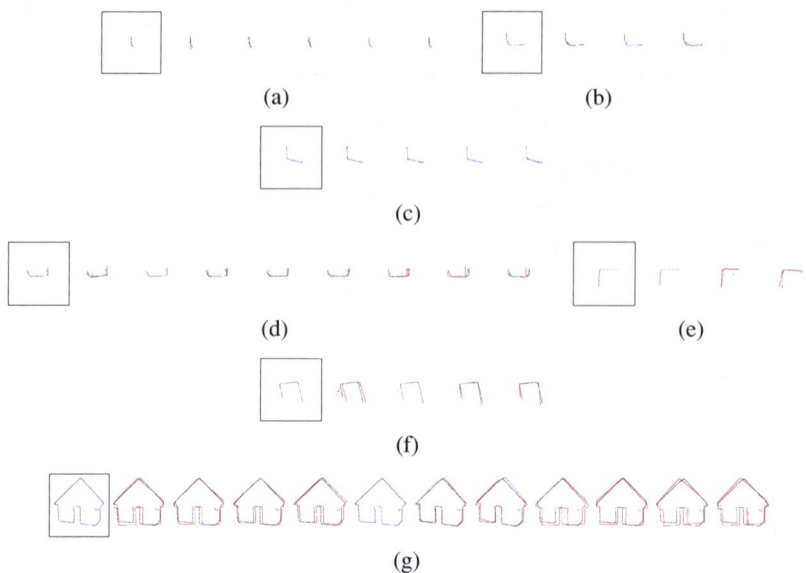

Fig. 5.16 Clusters of similar parts for object 51 (representative on the left in blue, cluster elements in red): (a) level 9, (b) level 11, (c) level 11, (d) level 12, (e) level 12, (f) level 13, and (g) level 17.

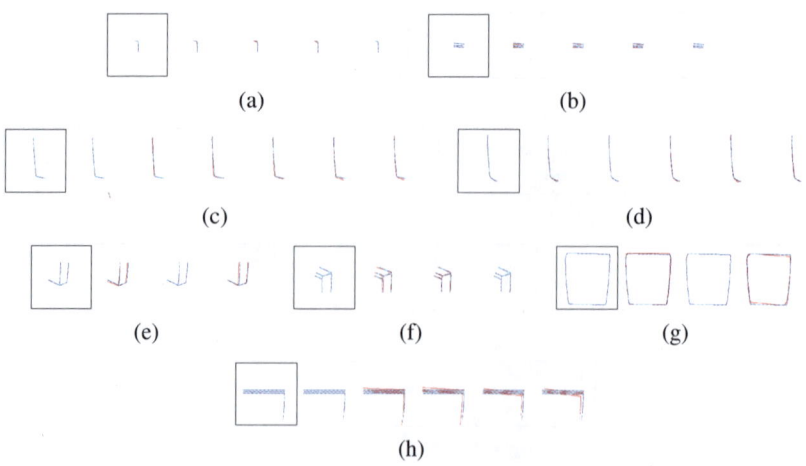

Fig. 5.17 Clusters of similar parts for object 65 (representative on the left in blue, cluster elements in red): (a) level 9, (b) level 12, (c) level 13, (d) level 13, (e) level 13, (f) level 14, (g) level 16, and (h) level 16.

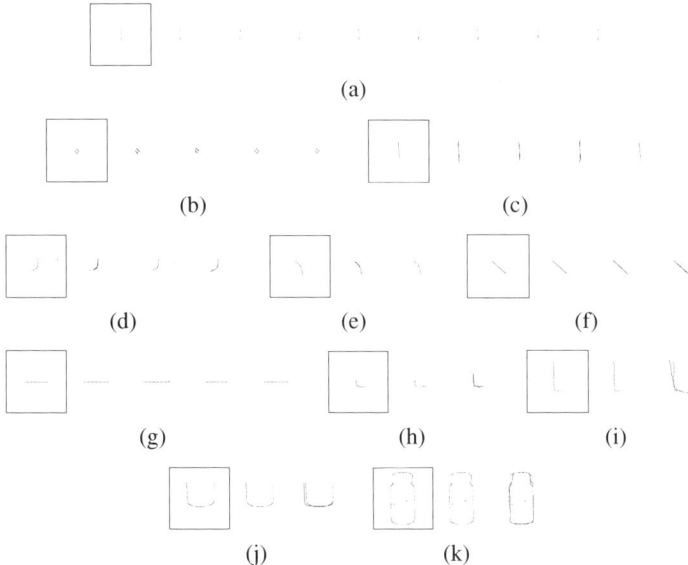

Fig. 5.18 Clusters of similar parts for object 1-100 (representative on the left in blue, cluster elements in red): (a) level 8, (b) level 9, (c) level 10, (d) level 10, (e) level 10, (f) level 11, (g) level 11, (h) level 11, (i) level 13, (j) level 14, and (k) level 16.

Table 5.5 Learning results for different online learning methods and the experimental setups: object 25, object 51, object 65 and object 1-100.

COIL object	me-thod	avg. similarity	self avg. simi-larity	num. nodes	num. edges	sharing degree	avg. num. shared nodes	log poste-rior
25	1	1.91	3.41	230	726	88.92	3.17	-133.72
25	2	45.41	25.33	184	728	88.68	3.90	-999.35
51	1	2.12	121.80	1009	3067	15.75	3.05	-449.95
51	2	11.71	126.82	437	2486	19.41	5.68	-526.58
65	1	1.36	213.44	1241	3791	17.16	3.08	-530.35
65	2	11.37	198.91	625	3417	19.04	5.46	-631.57
1-100	1	1.61	238.85	3325	9600	12.16	2.87	-1324.69
1-100	2	6.44	236.50	1984	9302	12.55	4.58	-1257.30

5.3.1.2   Online Learning

We applied the online learning framework proposed in Sec. 4.2.2 to the same examples as in Sec. 5.3.1.1 in order to make the results comparable to the offline learning results. For that, we learn a hierarchy based on the first element of each evaluation setup and sequentially add new training instances. For each example we apply the two sampling approaches for online learning:

- **Random top-down decomposition** using the state vector set (method 1). This method proceeds similar to the offline learning.
- **Top-down learning by recognition** (method 2). The already learned hierarchy is used to detect primitives and parts in the new training instance. The detected parts then allow to generate good segmentation hypotheses.

The results are summarized in Tab. 5.5. Surprisingly, the online learning reaches better log posterior values than the offline learning for object 25. Especially method 1 outperforms the offline learning and was able to generate good matching hierarchies with a reduced set of primitives. On the other hand, method 2 reaches worse log posterior values. They main reason for that is, that the distance function penalizes additionally large differences in the number of features. But method 2 skips remaining parts under a certain level causing missing features. However, method 2 leads to very small sets of nodes. The average number of parents for object 51 is 5.68, the largest value we found in our experiments. Fig. 5.19 shows the corresponding set of hierarchies for object 25. Compared to the set of hierarchies gathered during offline learning (see Fig. 5.5) the high log posterior value becomes obvious. Both sets of hierarchies, learned by method 1 (Fig. 5.19(a)) as well as by method 2 (Fig. 5.19(b)), seem to be more reasonable than the offline trained set. The main reason is that the search for an optimal decomposition of all primitives simultaneously has a quadratic complexity, while the sequential learning has just a linear complexity. Method 1 decomposes the cup spatially into three parts. It is noticeable that the parts are generated multiple times although they seem to be equal. Method 2 was able to describe all 72 root nodes with the same set of parts. Here, the silhouette and the edges of the opening were chosen. Especially at level 11 a few object-specific feature primitives were added. Due to its rotational symmetry object 25 is a special case. For all other training examples the online learning was not able to reach the same log posterior values as the offline learning. However, the log posterior is not that different for object 51: -428.47 (offline) against -449.95 (online), for object 65: -523.87 (offline) against -530.35 (online) and for object 1-100: -717.58 (offline) against -1324.69 (online). The overall conclusion is that online learning is especially suited for sets of similar instances, e.g. due to rotational symmetry. Here, online learning can efficiently apply and reuse already learned features. However, for multiple classes the offline learning is much more appropriate to find similar parts between objects of different classes.

5.3.2 Inference

In this section we investigate the inference mechanism as described in Sec. 3.8. Here, we (1) compare the standard straight bottom-up propagation with our new combined bottom-up and top-down propagation, (2) investigate the influence of the number of bottom-up sweeps and (3) investigate the influence of the number of level of details. In each experiment we use a set of varying features, parts and objects at

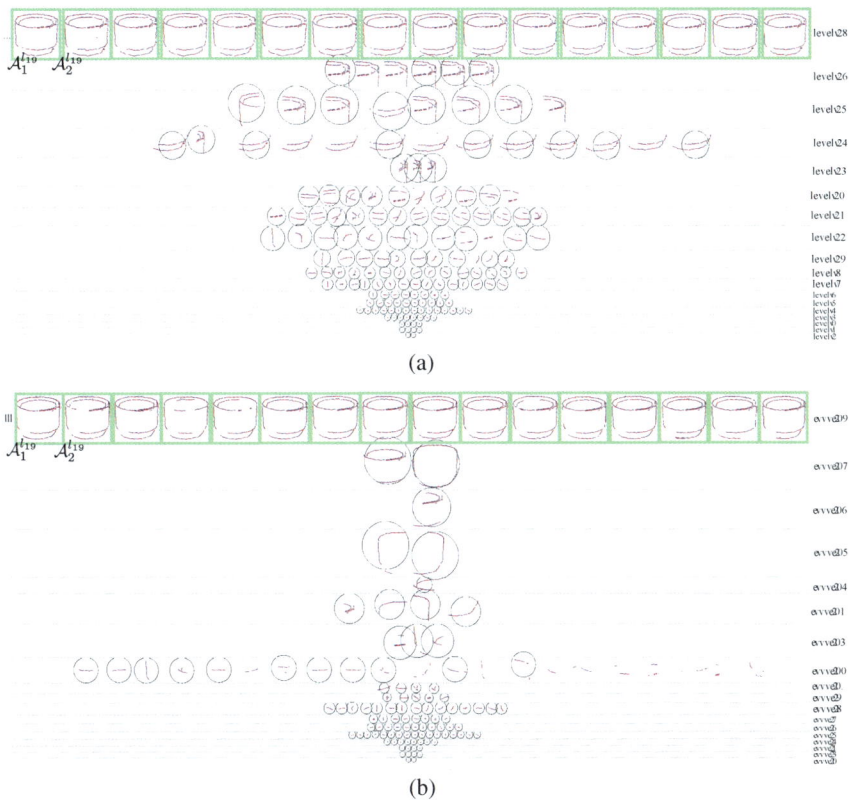

(a)

(b)

Fig. 5.19 Final online learned set of hierarchies for the representation of object 25 (root nodes are highlighted by a green bounding box). a) Random top-down decomposition (Method 1). b) Top-down learning by recognition (Method 2). Not all root nodes are shown.

different hierarchy levels to evaluate the performance. We obtained the corresponding ground truth data by using the learned hierarchies in a top-down manner. Nodes are randomly chosen and used to generate instances of its children. Given the position of the root node we generate iteratively instances of the children at the most likely position by taking the mean of the spatial relation. The position of the root node was determined by aligning the feature set of the evaluation instance to the feature set of the model. Detections within a σ-neighborhood (standard derivation of the spatial relation) to the ground truth position are taken as a correct detection, the others are regarded as false. As performance measures we use $recall = tp/(tp+fn)$ (true positive rate), $precision = tp/(tp+fp)$ (positive predictive value), root mean square error (RMSE) and the execution time.

5.3.2.1 Product Approximation

In order to compare the different sampling methods for product approximation we apply them on three challenging 2d examples. In each example we use the exact, importance, Gibbs, nearest neighbor, reduction/exact and reduction/nearest neighbor (see Sec. 3.8.1.2) sampler to draw up to 200 samples. In the first example, we had two mixtures with each 4 Gaussians (mixture 1 in Fig. 5.20(a), mixture 2 in Fig. 5.20(b), ground truth product density in Fig. 5.20(c)). We compare the products with the ground truth product density using the Kullback–Leibler (KL) divergence (as suggested in [98]). The results can be seen in Fig. 5.20(d). Especially the exact sampler has a slow convergence compared to the other samplers. The importance sampler and the reduction/nearest neighbor sampler produce accurate samples already within the first 30 iterations. The corresponding execution times are shown in Fig. 5.20(e). The reduction/nearest neighbor sampler has to apply the reduction method leading to an initial computational overhead. After this initial calculation, however, samples are generated with minimal additional costs. The second example contains two mixtures with each 40 Gaussians (mixture 1 in Fig. 5.21(a), mixture 2 in Fig. 5.21(b), ground truth product density in Fig. 5.21(c)). Here, the reduction/nearest neighbor sampler produces the most accurate samples, but as before the importance sampler is still faster. The third example is the one which is actually most related to the ap-

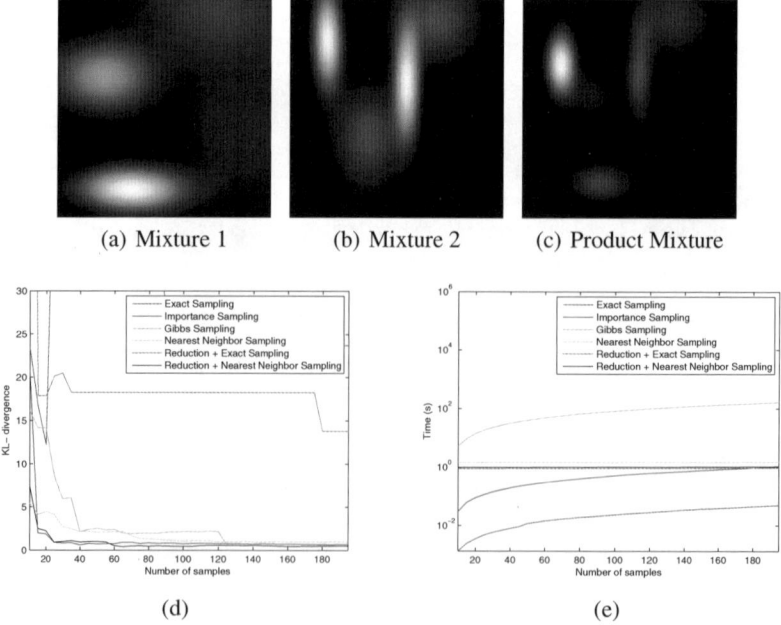

(a) Mixture 1   (b) Mixture 2   (c) Product Mixture

(d)   (e)

Fig. 5.20 Comparison of sampling accuracy and computation time for two mixtures with each 4 Gaussians

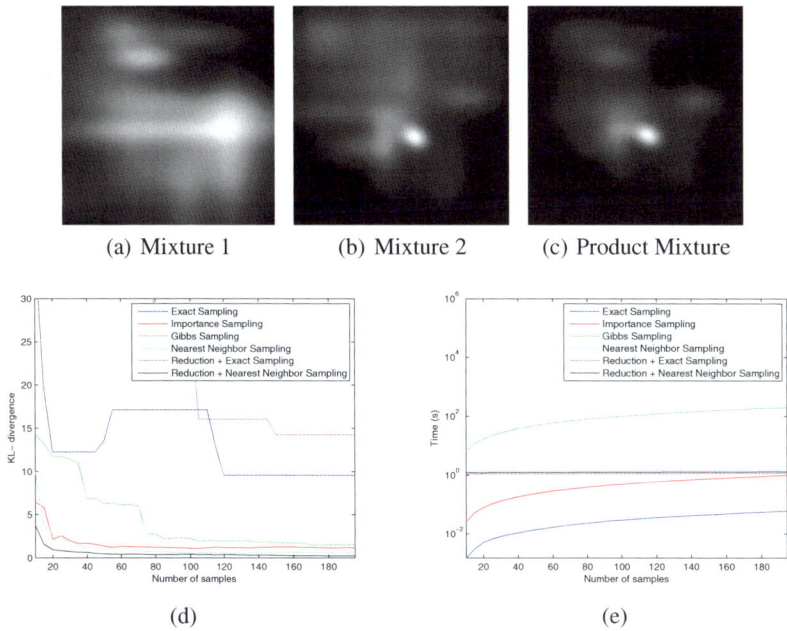

(a) Mixture 1   (b) Mixture 2   (c) Product Mixture

(d)   (e)

Fig. 5.21 Comparison of sampling accuracy and computation time for two mixtures with each 40 Gaussians

plications investigated in this monograph (mixture 1 in Fig. 5.22(a), mixture 2 in Fig. 5.22(b), ground truth product density in Fig. 5.22(c)). It is a density with multiple widely separated modes. For such a density, samplers like the Gibbs sampler are known to perform poorly [98]. However, the reduction/nearest neighbor sampler is much more accurate than the standard sampling approaches (Fig. 5.22(d)) and, furthermore, needs less computation time as can be see in Fig. 5.22(e).

5.3.2.2   Breath First vs. Depth First

A comparison of standard bottom-up propagation as e.g. used in [205, 66, 268] with our new combined bottom-up and top-down propagation can be seen in Fig. 5.23 -5.26. The bar diagram in Fig. 5.23 shows the average recall results plotted over the number of particles and the level. Without the top-down step (Fig. 5.23(a)) the recall performance increases with the number of particles at lower level ($\ell \leq 5$). However, higher levels ($\ell > 5$) show a worse performance. The good performance of the combined bottom-up and top-down propagation (Fig. 5.23(b)) is apparent. Even for 25 particles the approach shows better recall performance than the standard bottom-up propagation. A similar situation can be observed in Fig. 5.24, where the average precision is shown. For higher levels ($\ell > 5$) the true positive rate is zero, so that the positive predictive value has to be zero, too. The precision decreases with the level

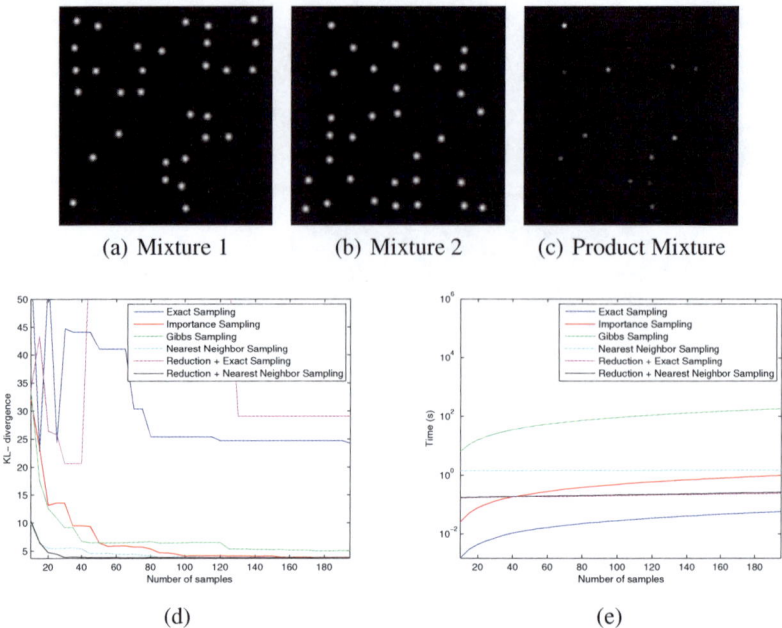

(a) Mixture 1   (b) Mixture 2   (c) Product Mixture

(d)   (e)

Fig. 5.22 Comparison of sampling accuracy and computation time for two mixtures with each 31 widely separated Gaussians

and reaches its minimum at level 7. One reason for that is the sharing which is very high at levels 5-8. Due to that high degree of sharing many particles are generated. At higher levels the precision increases again since the primitives becomes more object specific. The RMSE 5.25 seems to be independent of the number of particles. The general trend is a decreased RMSE for higher levels. Not surprisingly, the execution time increases with the number of particles as shown in Fig. 5.26. Since the number of elements increases with the level, the execution time increases with the level, too. Overall, the execution time is due to the additional top-down propagation step twice as large as without the top-down step.

5.3.2.3 Number of Sweeps

We now investigate the number of sweeps. Up to now, the particles were sent bottom-up in one sweep. Often it is better to send sets of particles in several sweeps (or passes) through the hierarchy. During the first passes mainly the high-weighted particles and the important nodes are chosen according to eq. 3.42 and 3.43. After a few sweeps the probability that low-weighted particles are chosen increases. In the first experiment we choose sets of two particles and in the second experiment sets of five particles. In Fig. 5.27 the bar diagrams over the number of sweeps and the level for the average recall are shown. For sets of two particles (Fig. 5.27(a)) the recall

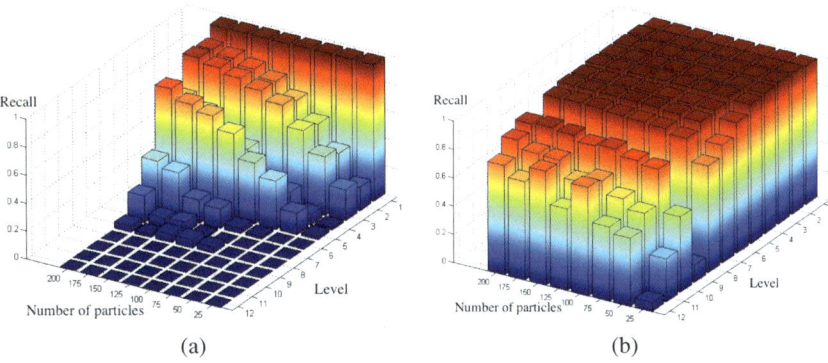

Fig. 5.23 Recall as a function of the number of samples and hierarchy level. (a) Straight bottom-up propagation. (b) Combined bottom-up and top-down propagation.

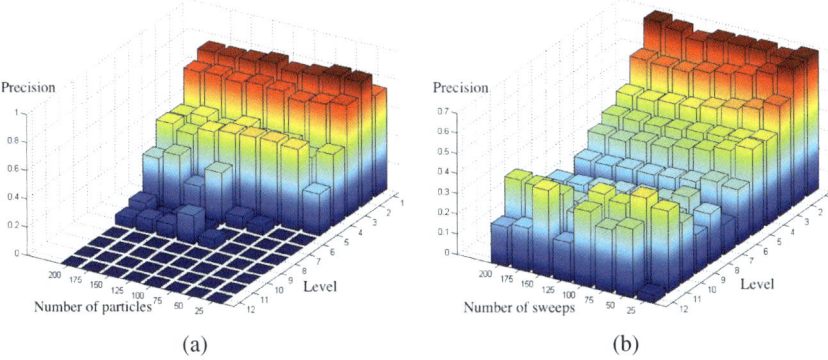

Fig. 5.24 Precision as a function of the number of samples and hierarchy level. (a) Straight bottom-up propagation. (b) Combined bottom-up and top-down propagation.

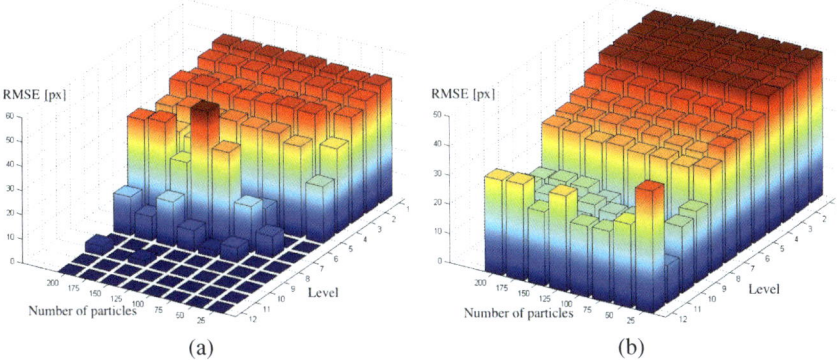

Fig. 5.25 RMSE as a function of the number of samples and hierarchy level. (a) Straight bottom-up propagation. (b) Combined bottom-up and top-down propagation.

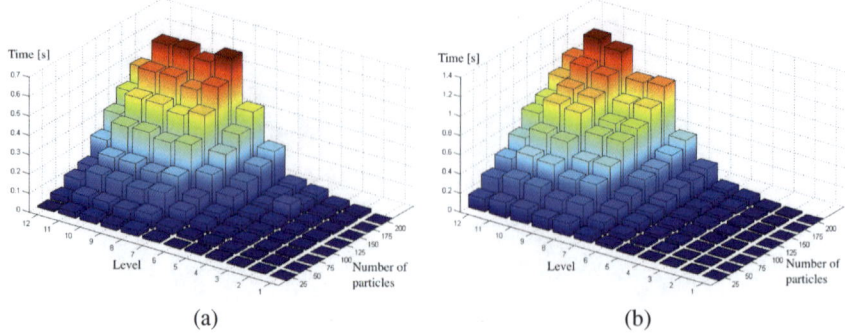

(a)   (b)

Fig. 5.26 Time as a function of the number of samples and hierarchy level. (a) Straight bottom-up propagation. (b) Combined bottom-up and top-down propagation.

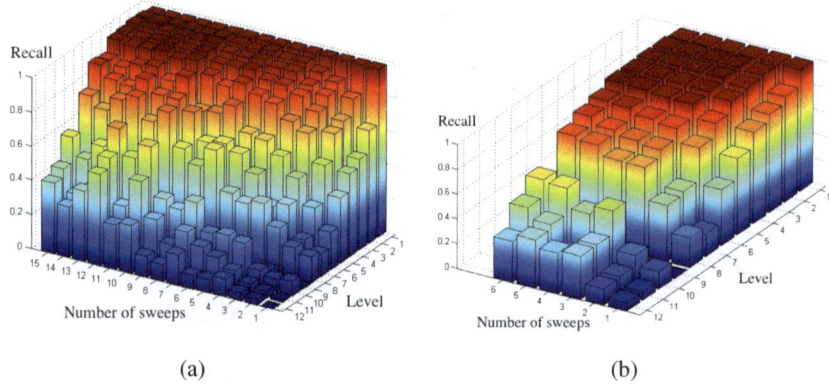

(a)   (b)

Fig. 5.27 Recall as a function of the number of sweeps and hierarchy level. (a) Bottom-up propagation of 2 particles per sweep. (b) Bottom-up propagation of 5 particles per sweep.

increases quickly during the first sweeps and runs into saturation. After 15 sweeps effectively 30 particles were propagated. The same number of particles were propagated in Fig. 5.27(b) after 6 iterations. The final recall performance is in both cases approximately equal. The same applies to the precision in Fig. 5.28 and the RMSE in Fig. 5.29. However, a crucial difference is the execution time (see Fig. 5.30). For the smaller set the overall execution time is approximately 2.5 times higher.

5.3.2.4 Coarse-to-Fine Inference

In this section we investigate the influence of the coarse-to-fine hierarchy as introduced in Sec. 3.7.4 on the performance. The learned hierarchies are extended to form a scale and rotation invariant representation. For that the hierarchy is shifted one level and connected to the scaled input images (see Sec. 4.2.5). The particles

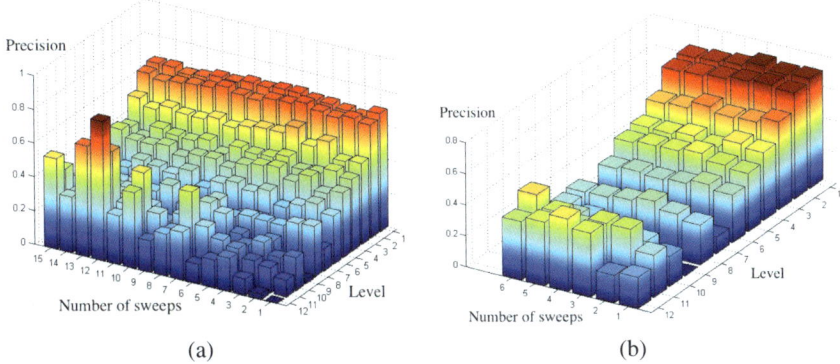

Fig. 5.28 Precision as a function of the number of sweeps and hierarchy level. (a) Bottom-up propagation of 2 particles per sweep. (b) Bottom-up propagation of 5 particles per sweep.

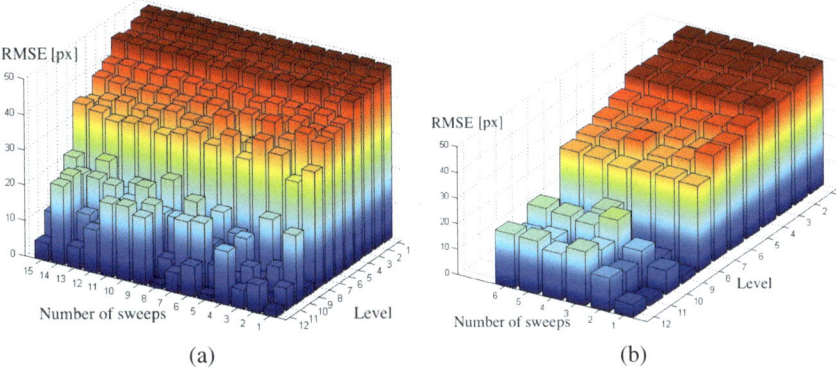

Fig. 5.29 RMSE as a function of the number of sweeps and hierarchy level. (a) bottom-up propagation of 2 particles per sweep. (b) bottom-up propagation of 5 particles per sweep.

are then injected into the hierarchy on different scale levels. The similarity hierarchy is constructed using the similarity threshold τ_s. Please note, that this allows one child to have multiple parent nodes. Results can be seen in Fig. 5.31. Fig. 5.31(a) shows examples for different classes. As can be seen, there is no similarity at higher levels of detail. At a coarse level of 4, similarity links between e.g. the 3 vehicles are established. In the case for multi-view instances is the similarity higher (Fig. 5.31(b)). Results for multi-scale and rotation can be found in Fig. 5.31(c) and Fig. 5.31(d). To measure the gain in performance when using the coarse-to-fine sampling scheme, we performed experiments with different numbers of similarity levels. Fig. 5.32 gives the results for different numbers of particles. Each bar diagram plots the recall over the level of detail and hierarchy level. A level of detail of 5 means that the similarity hierarchy has 5 levels. However, the ground truth data and the detec-

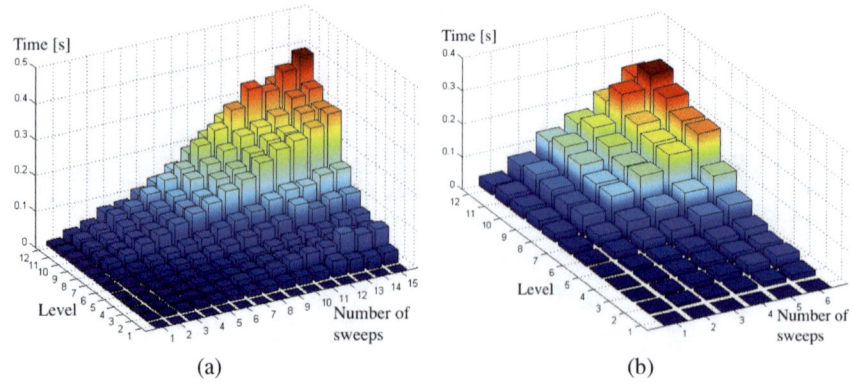

Fig. 5.30 Time as a function of the number of sweeps and hierarchy level. (a) Bottom-up propagation of 2 particles per sweep. (b) Bottom-up propagation of 5 particles per sweep.

tion results are still defined at the finest level 1. The similarity hierarchy is just used for better and more robust detection.

For 50 particles (see Fig. 5.32(a)) the advantage is already obvious. For 100 and 200 particles it becomes more and more apparent. Even for one additional layer the performance is significantly better. While the simple bottom-up propagation at the highest level of detail gets stuck in irrelevant hypotheses and thus is not able to detect the parts and objects at higher levels, the similarity hierarchies guarantee that the relevant information is efficiently propagated through the graph. A similar situation can be observed in Fig. 5.33, where the average precision is shown. Here, especially the high precision at higher hierarchy levels is noticeable. As before, the RMSE is reduced at higher hierarchy levels (as can be seen in Fig. 5.34). Fig. 5.35 shows a plot of the execution times. Each additional level of detail significantly reduces the execution time since the bottom-up propagation is initiated at a coarser resolution. The main reasons are that (i) the number of low-level hypotheses detected in the input image is reduced, and (ii) the number of primitives in the hierarchy is reduced. The time needed for the top-down evaluation in the similarity hierarchy can be neglected.

Detection examples are shown in Fig. 5.36, Fig. 5.37 and Fig. 5.38. Fig. 5.36(a) shows the input image and Fig. 5.36(b) the detection results for a simple bottom-up propagation for object 65. Low-level hypotheses are generated at several object positions, however, the approach fails to detect object 65 correctly. In Fig. 5.36(c) the result of the coarse-to-fine approach is shown. The object was correctly detected in the cluttered image. Both results were achieved with the same number of particles. In the test image in Fig. 5.37(a) we added Gaussian noise to the color values. 5.37 shows the result when using just simple bottom-up propagation. While it again fails, the use of the coarse-to-fine propagation was able to detect the two tall versions of the object 65 (see Fig. 5.37(c)). Fig. 5.38(b) shows the result for the coarse-to-fine propagation without Gaussian noise. All objects were correctly detected at different

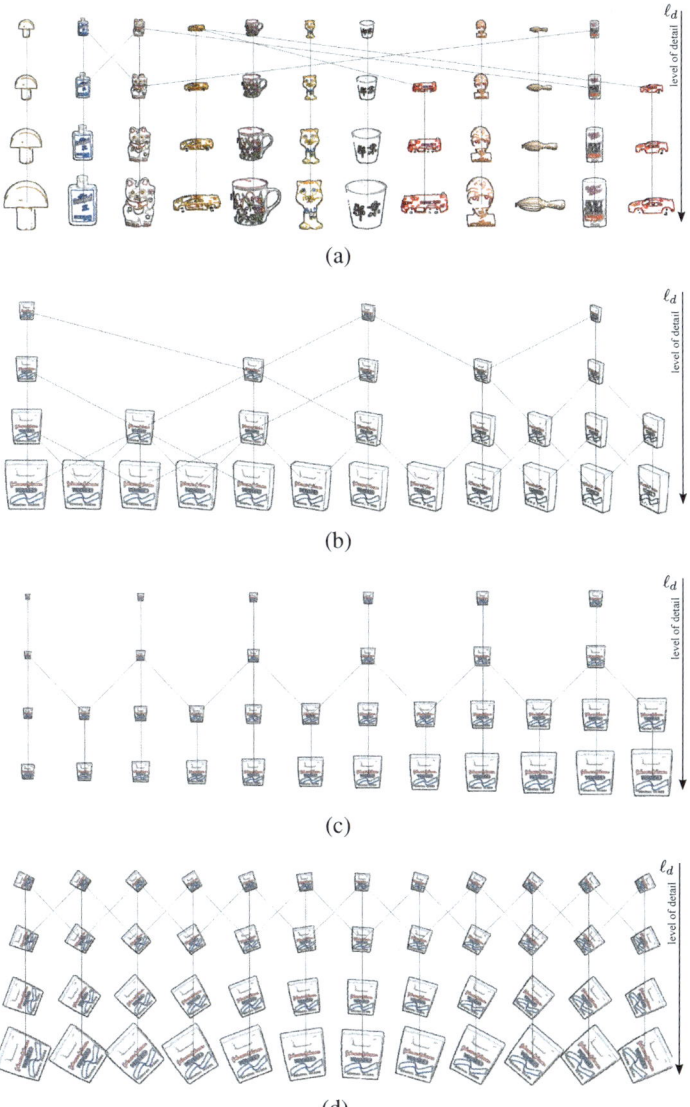

Fig. 5.31 Similarity hierarchies for: (a) multi classes, (b) multi view, (c) multi scale, and (d) multi orientation. All hierarchies are obtained using the same clustering approach and the same parameters.

scales. In Fig. 5.38(d) the object was occluded by other objects. As can be seen, two objects were correctly detected although parts are occluded. The remaining objects were not detected. However, at least their parts are correctly classified during top-down propagation. An example for the recognition of multiple objects at multiple views and scales can be seen in Fig. 5.38(f).

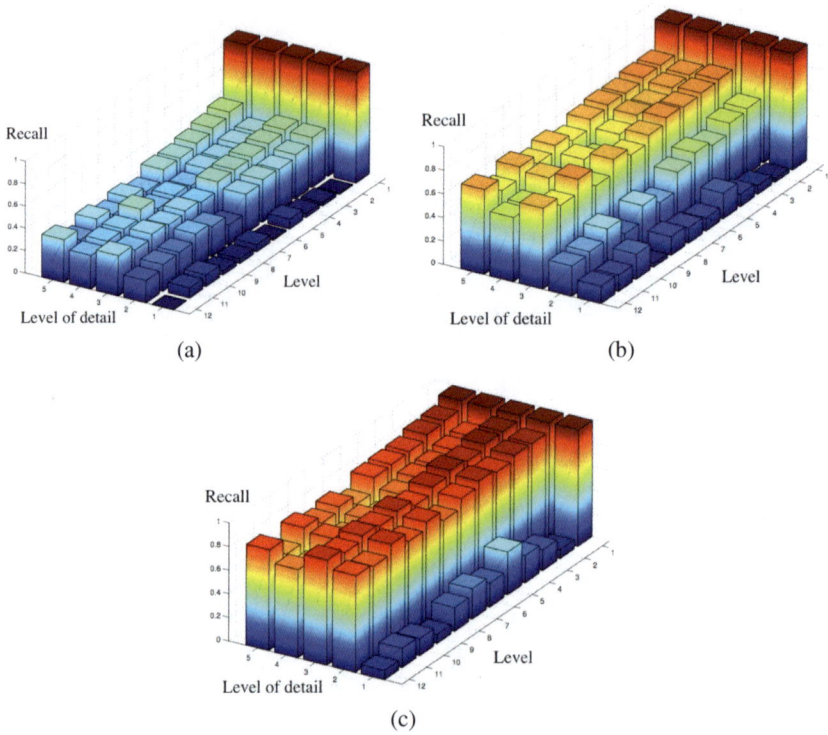

Fig. 5.32 Recall as a function of the level of detail and hierarchy level. (a) 50 particles. (b) 100 particles. (c) 200 particles.

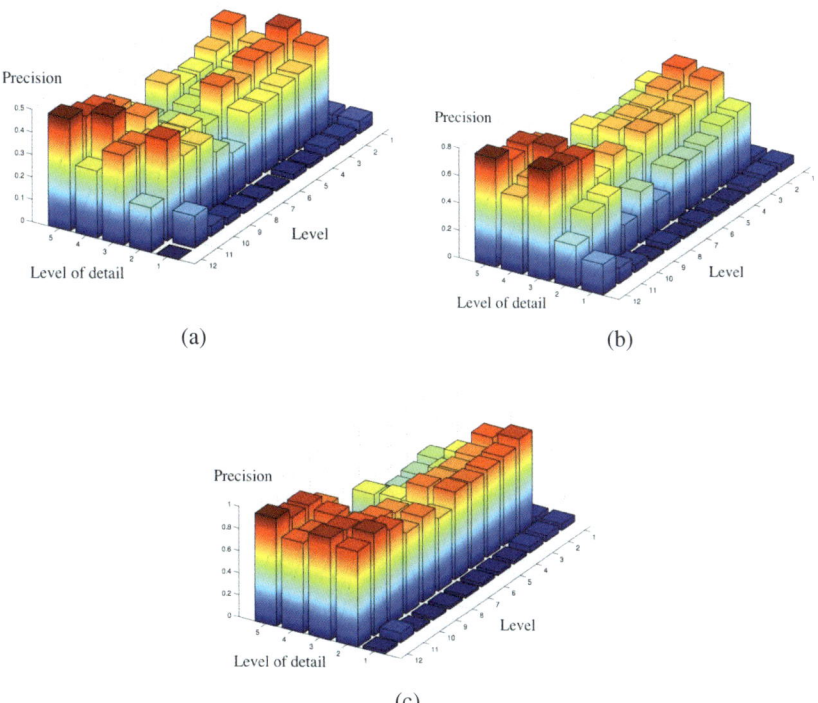

Fig. 5.33 Precision as a function of the level of detail and hierarchy level. (a) 50 particles. (b) 100 particles. (c) 200 particles.

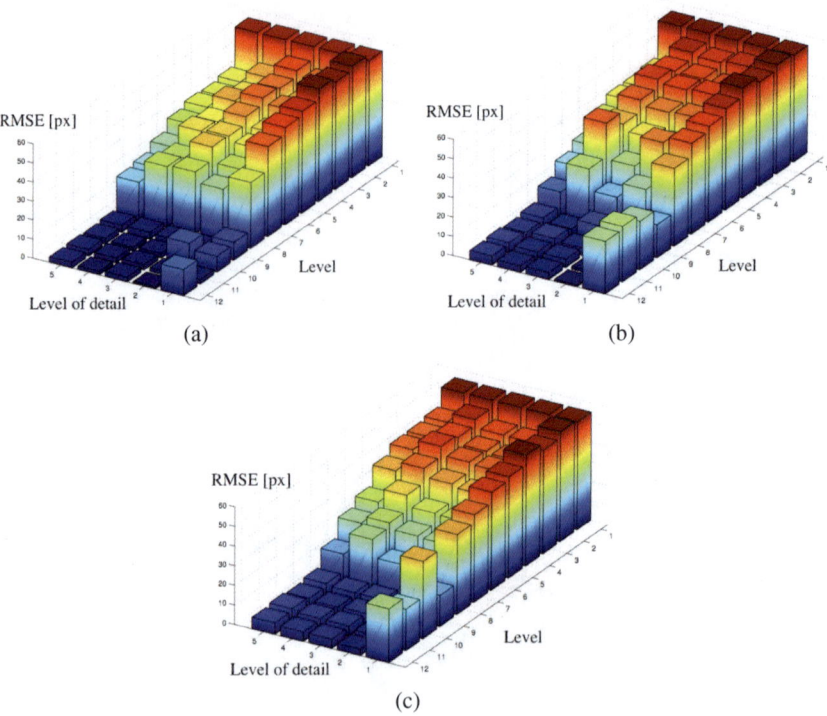

Fig. 5.34 RMSE as a function of the level of detail and hierarchy level. (a) 50 particles. (b) 100 particles. (c) 200 particles.

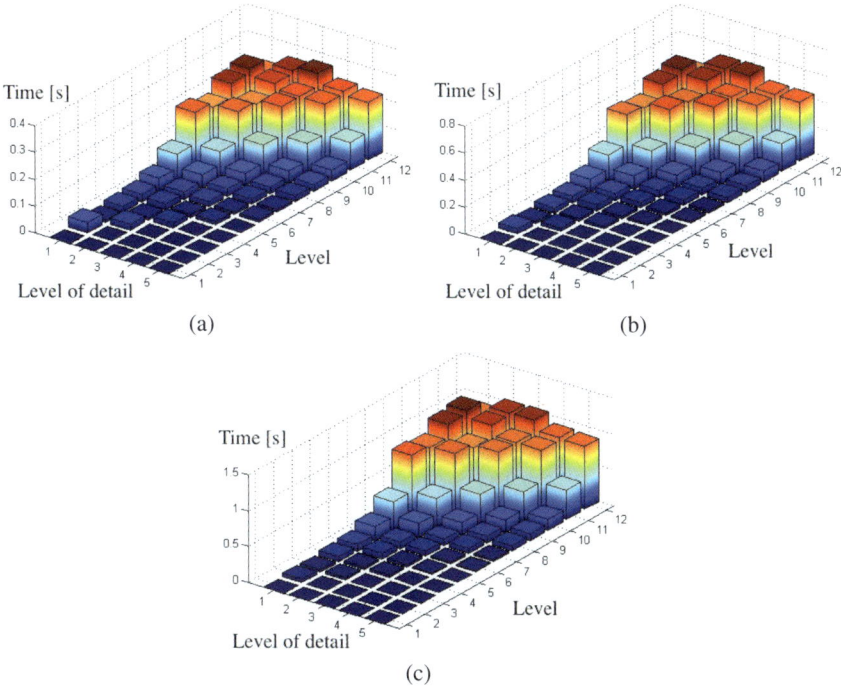

Fig. 5.35 Time as a function of the level of detail and hierarchy level. (a) 50 particles. (b) 100 particles. (c) 200 particles.

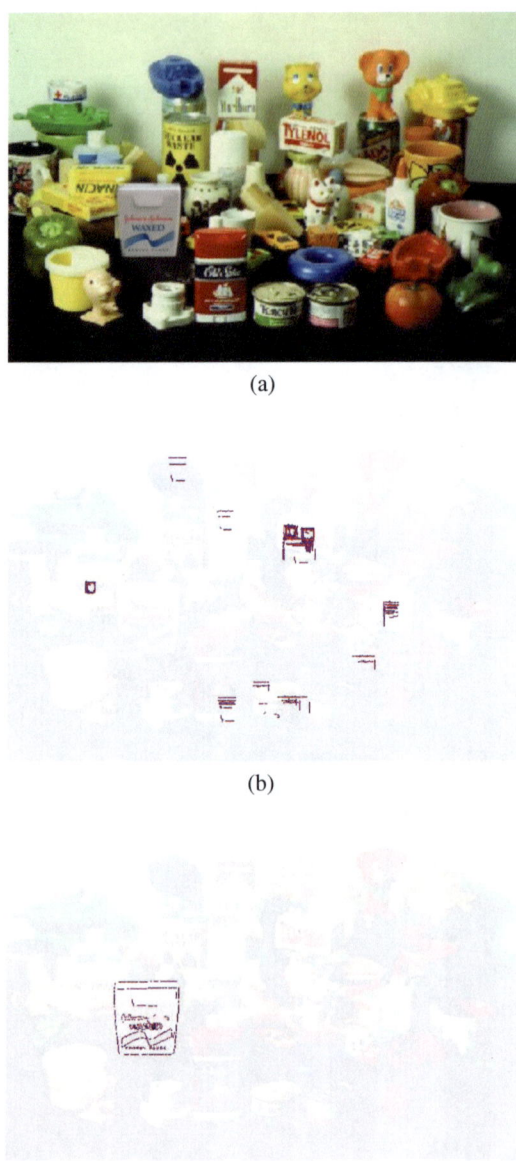

(a)

(b)

(c)

Fig. 5.36 Recognition example of object 65 with cluttered background: (a) input image, (b) without similarity hierarchy, and (c) with similarity hierarchy and five levels of detail.

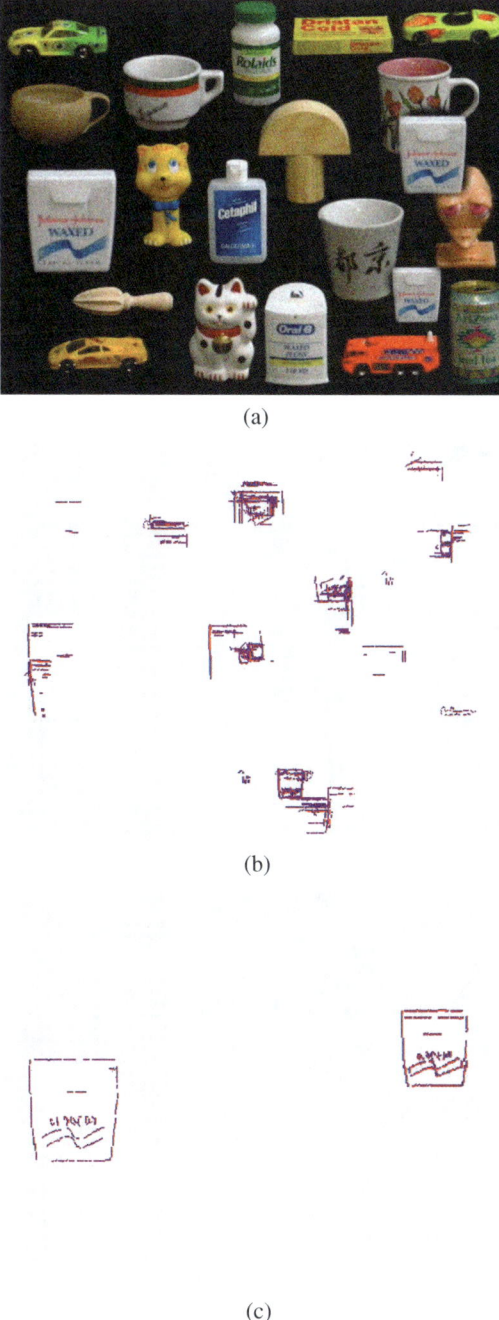

(a)

(b)

(c)

Fig. 5.37 Recognition example of object 65 with added Gaussian noise: (a) input image, (b) without similarity hierarchy, and (c) with similarity hierarchy and five levels of detail.

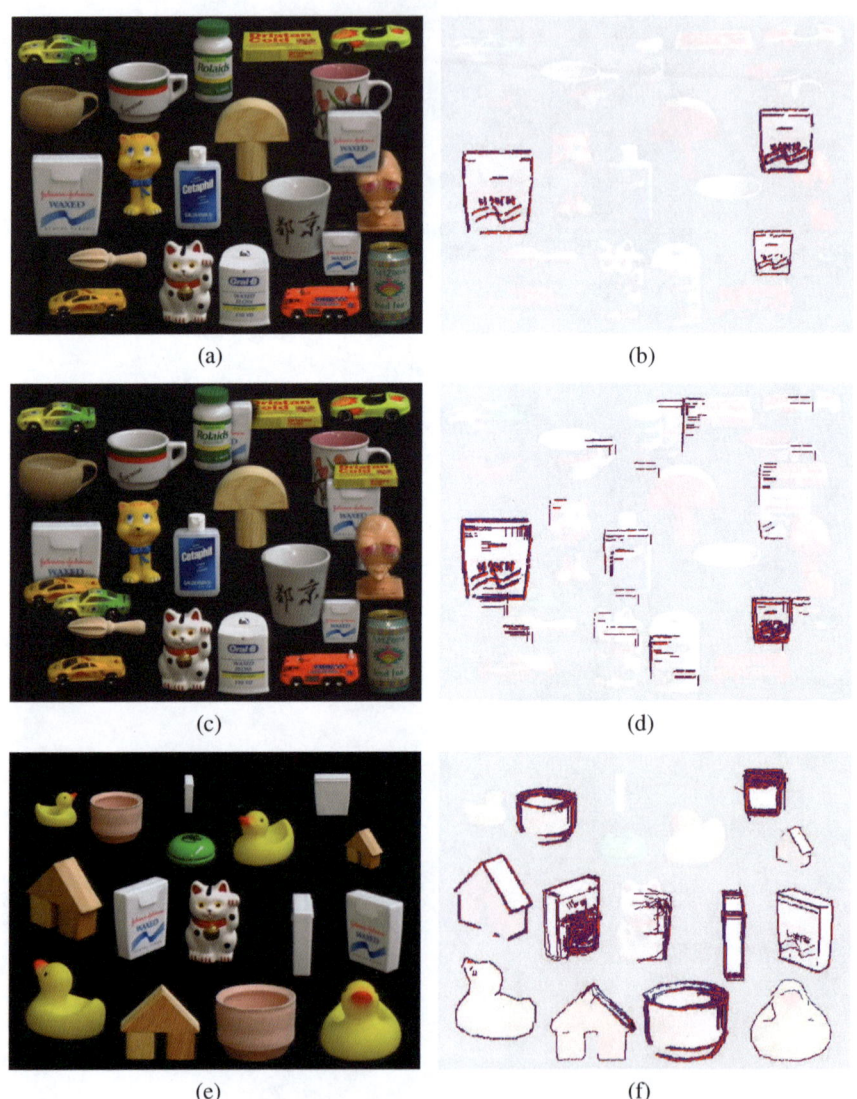

(a)   (b)

(c)   (d)

(e)   (f)

Fig. 5.38 Recognition examples: (a) input image and (b) recognition result of object 65 at different scales, (c) input image and (d) recognition result of object 65 with occlusions, (e) input image and (f) recognition result of multiple objects.

Chapter 6
Human Pose Estimation

The aim of human pose estimation is to detect and estimate the configuration of the articulation structure of a person. Human pose estimation has become a popular research topic including a wide range of approaches. Marker-based approaches, for example, are one of the most precise methods, but they need special markers attached to the human body. In this work we concentrate on monocular vision-based approaches, which are not using additional means like markers. We are also not using other common methods like foreground–background segmentation or temporal tracking in order to guarantee that our approach is independent of the quality of such methods.

However, especially tracking approaches are common for human pose estimation, and therefore we will briefly describe the basic idea and how tracking information can be integrated into our representation. From a Bayesian perspective, we are interested in some degree of belief in the state x_t at time t, given measurement y_t. This belief can be expressed by the PDF $p(x_t|y_t)$ and the belief can be calculated using Bayes' rule as

$$p(x_t|y_t) \propto p(y_t|x_t)p(x_t) \tag{6.1}$$

The likelihood $p(y_t|x_t)$ is in general a probability density from which it is difficult to directly draw samples from. This is the reason why, e.g. in tracking approaches, one tries to estimate $p(x_t)$ using the previous time steps. The prior is estimated using $p(x_t) \propto p(x_t|y_{1:t-1})$ and is obtained via $p(x_t|y_{1:t-1}) = \int p(x_t|x_{t-1})p(x_{t-1}|y_{1:t-1})dx_{t-1}$. In a belief propagation framework, this prior would correspond to the message from node x_{t-1} to node x_t. It is a prediction of the new state at time step t based on all measurements of the last time steps $y_{1:t-1}$. Here, $p(x_t|x_{t-1})$ models the dynamic properties, and $p(x_{t-1}|y_{1:t-1})$ is the belief of the state at the last time step. In the update stage, the measurement is used to modify the prior density to obtain the required posterior density of the current state. This is the key idea of many common tracking approaches [195, 214, 124]. It avoids the difficult problem of evaluating the whole likelihood density by restricting the evaluation based on the prediction of the prior. Unfortunately, this main advantage

© Springer International Publishing Switzerland 2015   121
J. Spehr, *On Hierarchical Models for Visual Recognition & Learning of Objects, Scenes, & Activities*,
Studies in Systems, Decision and Control 11, DOI: 10.1007/978-3-319-11325-8_6

is simultaneously a large drawback, since the restriction can cause the tracking to fail, if e.g. the dynamic model is too simple or the number of particles is too small. In addition, tracking approaches need an initialization of the human pose which is often not given in real world scenes.

The integration of tracking information into our hierarchal representation is straightforward. As already mentioned in Sec. 3.4, we just have to add the information from the temporal prediction by means of an additional factor in eq. 3.2. Related to eq. 6.1 we thus combine the likelihood of a node $p(y_t|x_t)$ (provided by the hierarchical structure) with the prior $p(x_t)$ (estimated by the prediction).

In this work we focus on an efficient evaluation of the likelihood function without using prediction from the previous time steps. From an analysis-by-synthesis point of view our motivation is: for every state candidate the human appearance has to be synthesized and compared with the image features. But many candidates are quite similar and often candidates are even sharing subparts. Thus, using similarity and avoiding redundant evaluations are the key issues of our novel hierarchical representation.

The rest of the chapter is organized as follows: Sec.6.1 will briefly summarize related works. Sec. 6.2 explains the idea of scalable hierarchical human pose representation and motivates the sharing of visual primitives. A semi-supervised learning approach is presented in Sec. 6.2.1. Experimental results are presented and discussed in Sec. 6.3 using a gait analysis application. A preliminary version of this chapter was published in [223].

6.1 Related Work

In the following we will briefly summarize related works. A more general overview of vision-based human motion capture and analysis can be found e.g. in [149, 150].

6.1.1 Human Pose Estimation

Human pose estimation is challenging mainly due to the many degrees of freedom of the human body. Model-based generative approaches are using the 3d geometric representation of human shape and kinematic structure to reconstruct poses in an analysis-by-synthesis manner. In general, these approaches try to optimize the similarity between the projected model and the observed image information [89, 238]. A common assumption is, that an initialization of the human pose in the first frame of a sequence is given. Thus, the pose estimation can be regarded as a tracking problem. Rohr [195] used in his work a matching approach to compare the edges of a projection of a 3d body model to those found in the image.

Other approaches which are related to our work are exemplar-based methods, where the human body is represented by a set of 2d views. Mori and Malik[155] stored 2d views of the human body in different configurations and viewpoints, and manually marked the locations of the body joints. During recognition, shape context matching was used to match the new shape against the training data. A shape matching algorithm was also used by Sullivan and Carlsson [229]. They recognized and tracked human actions based on exemplars, which were key frames in an image training sequence. Since the number of examples can become high, Shakhnarovich et al. [212] learned a set of hashing functions in order to efficiently index the examples during recognition.

Most related to our work are part-based models, where the human body is represented by a constellation of connected body parts like head, torso, legs. The main two components of these models are body part detectors and configuration models, which define how parts are spatially arranged to each other. A wide range of specific part detectors has been employed, which are based on 2d shapes [193], svm classifiers [196], poselets [21], "shouters" [214], multi-view eigenspaces [214], skin color [124, 94, 263], AdaBoost [147, 124], pairs of parallel line segments [99, 189], segmentation-based [156, 154, 224], HOG features [56], or locally initialized appearance models [186]. Ju et al. [105] proposed a person model, which represents the human body by a set of connected planar patches. The so-called "cardboard person model" uses the chain structure to facilitate the estimation of articulated motions. Felzenszwalb and Huttenlocher [57] proposed pictorial structures, where the parts encode the local appearance of the object, and spring-like connections between pairs of parts are used to represent deformable configurations. The matching of a pictorial structure to an image is formulated as a minimization problem of an energy function, which encodes the cost of part matching with the image data, and the cost of the relative locations of the parts. Ioffe and Forsyth [99] represented the human body by an assembly of 9 rectangular segments. A rectangle detector delivers a set of candidates which are assembled by sampling based on kinematic constraints. Segmentation-based pose estimation like [156, 154] uses segmentation to guide the algorithms to salient regions of the image. Mori et al. [156] built a limb and torso detector based on the segmentation. Srinivasan and Shi [224] used segmentation as well. They evaluated partial body masks via shape matching with exemplars. Sigal et al. [214] used nonparametric belief propagation [227] for human pose estimation. The inference method is applied to a loose-limbed graphical model, where each node represents one body part. Ramanan et al. [186, 185] proposed a person detector that can detect and localize limbs of people. In order to increase the power of the features they proposed an iterative parsing process, where better and better features tuned to a particular image are sequentially learned. Bourdev and Malik [21, 20] used a simple Hough voting scheme for person detection. Poselets describe parts of a human's pose; they are used to vote for the center of the person. Coarse-to-fine hierarchies are e.g. proposed by Gavrila [75], who used a hierarchical template matching to detect pedestrians from a moving vehicle. A coarse-to-fine cascade of pictorial structure models was proposed by Sapp et al. [201]. Similar to the popular Viola-Jones classifier [237], they trained a cascade of structured models to refine the set of candidates iteratively level by level. At each level they pruned as much candidates as possible while preserving true candidates. A combination of poselets and the coarse-to-fine cascade was proposed in [244].

The main contributions of our novel hierarchical human pose estimation are: (1) We propose a compositional hierarchy, which allows to estimate efficiently the pose of the human body by means of sharing primitives. The whole algorithm is integrated in an unified probabilistic framework. (2) The hierarchical representation provides a scalable model of the human body with varying level of detail. (3) We demonstrate the approach in a gait analysis application.

6.1.2 Human Gait Analysis

We will now give a brief overview of work related to human gait analysis. Human gait analysis is of particular interest for the identification of a human. Analyses based on the global body structure and the global body dynamics are discussed in many publications. Wang et al. [242] tried to identify humans using a spatial-temporal silhouette analysis. This analysis implicitly captures the structural and transitional characteristics of gait based on a eigenspace transformation, which is applied to time varying distance signals derived from a sequence of silhouette images. Kale et al. [108] defined a HMM to model the dynamics of individual gait. The HMM has five hidden states which are associated to five representative binary silhouettes. For each individual a HMM is generated in the training phase. During the recognition phase, the HMM with the largest probability identifies the individual. Yam et al. [256] defined a gait signature based on Fourier analysis of the variation in the motion of the thigh and lower leg. The leg motion is extracted by temporal template matching with a model defined by forced coupled oscillators as basis. Davis and Gao [38] recognized the gender of a person based on the gait. Trajectories from motion capture devices of prototype female and male walkers are factored using three-mode PCA into components interpreted as posture, time, and gender.

6.2 Hierarchical Human Pose Representation

Consider e.g. a frontal image of a person (Fig. 6.1 (left)). Then, the whole body is e.g. composed of the upper body, the torso with arms and the legs. The upper body is, in turn, composed of the head and the shoulders. And finally the head is composed of visual primitives like parallel lines, corners, arcs and so on. Without sharing, for each configuration the whole composition has to be compared to the measurement. Unfortunately, the configuration space is quite large (Fig. 6.1 (top, right) shows just a few examples of configurations). However, by using a hierarchical decomposition the space can be reduced at each lower level. It can be seen in Fig. 6.1 (right), that at lowest level just a few configurations of gradients have to be evaluated. The evaluation results can be combined to more complex visual primitives on the next higher hierarchy level and so on. Of course the number of configurations at the top is still large. However, while the number of configurations increases, the number of likely hypotheses decreases with each level.

The hierarchical model corresponds with few exceptions to that used in Chap. 5 for object pose estimation. The model is therefore specified by a set of hierarchies $\mathcal{G} = \{\mathcal{G}_1, ..., \mathcal{G}_N\}$, which are undirected tree-structured graphs $\mathcal{G} = (\mathcal{V}, \mathcal{E})$ (see Sec. 3.7). The set of nodes $\mathcal{V}$ represents low-level features (leaf nodes), body

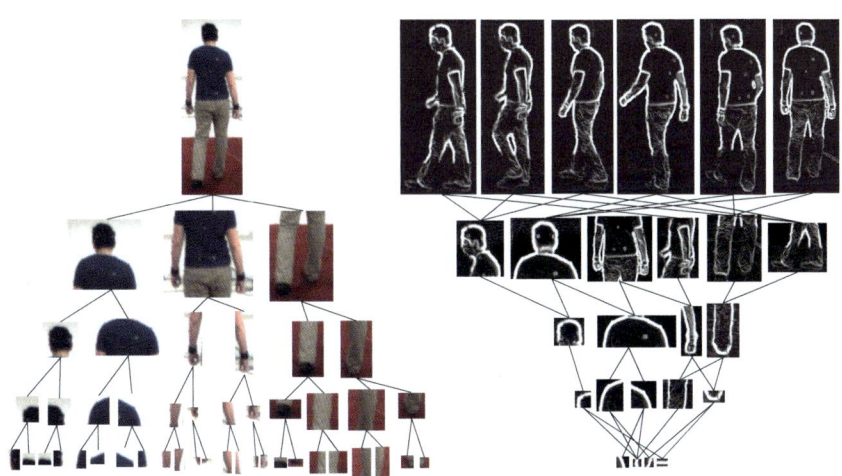

Fig. 6.1 Example of the hierarchical decomposition of the human body. The human body is composed of body parts, and body parts are, in turn, composed of visual primitives (left). Instead of evaluating all configuration of the human body separately we take advantage of the circumstance that body parts share visual primitives (right).

parts and the whole body (root nodes). Each configuration of the human body is represented by one hierarchy with one root node. The corresponding set of edges $\mathcal{E}$ define 2d relations between the parts. As before, we consider hierarchical graphical models, where just the low-level nodes are associated with a noisy local observation. The only difference is that we are now using discrete random variables (see Sec. 2.2) and are therefore not applying the nonparametric belief propagation framework. Instead each PDF is represented by a 2d probability map. We apply the standard BP algorithm as described in Sec. 2.3.1. Since our pairwise potential functions $\psi_{ji}(x_i, x_j)$ only depend on the difference between its arguments $\psi_{ji}(x_j, x_i) = \psi_{ji}(x_j - x_i)$ we can calculate messages by discrete convolution as described in eq. 2.27. Our convolution kernels are defined by a simple Gaussian $\psi_j(x_i, x_j = \tilde{x}_j^{(\ell)}) = \mathcal{N}(x_i; (\tilde{u}_j^{(\ell)}, \tilde{v}_j^{(\ell)}) + \boldsymbol{r}_{ji}, \Lambda_i)$ with a relative position vector $\boldsymbol{r}_{ji}$ (see Sec.3.6 for a detailed description). Λ_i is a diagonal covariance matrix $\Lambda_i = \sigma_{\ell_i} I$ with $\sigma_{\ell_i} = \alpha_\sigma (\beta_\sigma)^{\ell_i}$ (see Sec. 4.2.3). The increasing values of σ_{ℓ_i} at higher levels mean simultaneously an increasing kernel size leading to an additional computational effort at higher hierarchy levels ℓ_i. In order to avoid this problem we downsample the size of the probability maps level by level according to a scaling factor s. The size of the probability maps thus corresponds to the original image size at level 1, and is downsampled by s^{-1} for each higher level. This scaling allows us to use the same convolution kernel at all hierarchy levels. Since a precise localization of the whole human body is not necessary in our case, the rough localization in the low resolution probability map is sufficient.

Fig. 6.2 Samples from the human pose dataset

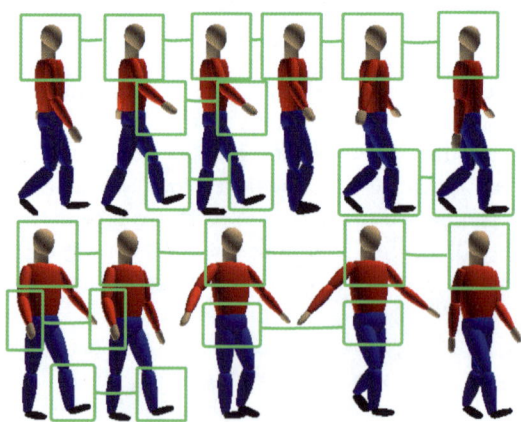

Fig. 6.3 Samples from the human pose dataset. The green bounding boxes visualize human body parts, which can be shared between different configurations.

6.2.1 Learning of the Hierarchies

We will now describe, how to learn an optimal representation of a set of n configuration instances $\mathcal{O} = \{\mathcal{O}_1, ..., \mathcal{O}_n\}$. For that, we use the online learning (see Sec. 4.2.2), and use the Metropolis sampler to get samples from the posterior distribution $p(\mathcal{G}, \theta_\mathcal{G}|\mathcal{O})$. The final set of hierarchies corresponds to that sample, which maximizes $p(\mathcal{G}, \theta_\mathcal{G}|\mathcal{O})$. The configuration instances $\mathcal{O} = \{\mathcal{O}_1, ..., \mathcal{O}_n\}$ where synthetically generated. Samples are shown in Fig. 6.2 and the idea of sharing is illustrated in Fig. 6.3.

We manually guide the learning by sequentially adding particular object shape classes. We start with line elements which are added at level 8 and are composed of about 300 low-level features. We learn the scale and rotation invariant representation of the line shape class according to Sec. 4.2.4. The hierarchical decomposition is therefore just learned once and the building rule is shared among all scaled and rotated instances. After the hierarchical representation has learned lines we manually add arc shapes, parallel lines, and head shapes as depicted in Fig. 6.5. The shape primitives are then used to describe the body parts. Finally, the body parts are used to describe the whole human body. As result we get a more comprehensible hierarchical decomposition of the human body in terms of geometric primitives. Fig. 6.5 gives an overview of our hierarchy and the used primitives.

The hierarchical structure makes use of two different sharing principles. On the one hand sharing of primitives and basic shapes allows an efficient inference scheme. On the other hand sharing of building rules can accelerate the learning procedure and furthermore provide a consistent hierarchical structure. The building rules as depicted by arrows in Fig. 6.5 always refer to the same shape classes. Just the relative position vectors r_{ji} are scaled and rotated. Vertically, we place the shape classes according to the level which also represents the scale of the parts and objects. Horizontally, the classes are placed according to their complexity. Simple classes like the lines already exist at level ℓ_1. The more complex a class structure becomes the higher is its first scale in the hierarchy (ℓ_6 for the whole human body). Please note that the orientations are roughly quantized at lower levels and are more accurately quantized at higher levels. At level ℓ_1 we are using simple gradient information in just four gradient directions. At higher levels the direction and size information gets more and more accurate. This is a consequence of the constant threshold of the sharing criterion. Since the similarity increases as the lines are downsized the number of shared lines increases as well.

6.3 Experimental Results

Human pose estimation is of particular interest for human gait analysis. The localization and detection of the human pose including body parts, especially the head and feet, is very important for the extraction of gait parameters like step width, step height, speed and speed variance, gait harmony, compensation motion and body sway. In medical telemonitoring applications these parameters can be used to allow a remote diagnosis. While the patient is living in her/his home environment, the telemonitoring system is extracting vital signs of the patient and a medical practitioner can use these for diagnosis. Thus, especially elderly people can live a longer time in their familiar environment, while simultaneously their health is monitored. One of the most challenging steps in this scenario is the extraction of the gait parameters. There are numerous vision-based approaches, which can be used for this task. Unfortunately, most of them are not appropriate for use in home environments.

We assume a camera setup, where the human body is captured from a frontal view. The main advantage is that this setup can be easily realized in standard home environments. Unfortunately, this makes the human pose estimation challenging since it requires scalable human pose estimation. When the patient is approaching the camera the overall size of the human body in the camera image increases. Therefore the pose estimation has to deal with a changing scale. Fortunately, our hierarchical representation offers intrinsically a scalable representation and is thus ideally suited for this kind of application.

(a)   (b)

Fig. 6.4 Positive and negative evaluation examples of the human body. (a) Positive and (b) negative examples.

In our work, we are interested in human gait parameters like step width, compensation motion or body sway. In order to be able to extract these parameters one has to detect the person and the body parts within the image first. Then, a calibrated camera setup allows to reconstruct the positions of the body parts in 3d using additional constraints like e.g. contact to the ground plane. These 3d positions can be used to estimate human gait parameters. We decompose the human body into primitives e.g. body parts like head, arms, torso and legs. The body parts are, in turn, decomposed into visual primitives like geometric shapes. In order to demonstrate the proposed hierarchical framework we built up a gait database of 36 patients. It contains video sequences with patients approaching and departing the camera from a frontal perspective. In our experiments we applied the human body hierarchy (Sec. 6.2.1) to the images. Fig. 6.8 shows image samples for one of the sequences. The images contain the patient departing from the camera and the therapist. The recognition results are highlighted in red (head and torso), blue (left leg), and green (right leg). One can see, that the hierarchical representation is able to detect the person at different scales without explicit scaling of the input image. Fig. 6.9 shows results for other patients. Although the subjects differ strongly in their appearance e.g. one subject wears tight-fitting clothing, the hierarchical structure ensures the overall appearance while locally the body parts and visual primitives are just loosely connected and thus allow the adaption to the local observation. Difficulties arise with walking frames, as can be seen in Fig. 6.9 (top, right), or with textured clothing Fig. 6.9 (bottom).

The corresponding evaluation results can be seen in Fig. 6.6. The evaluation dataset contains positive and negative examples of the human body (examples

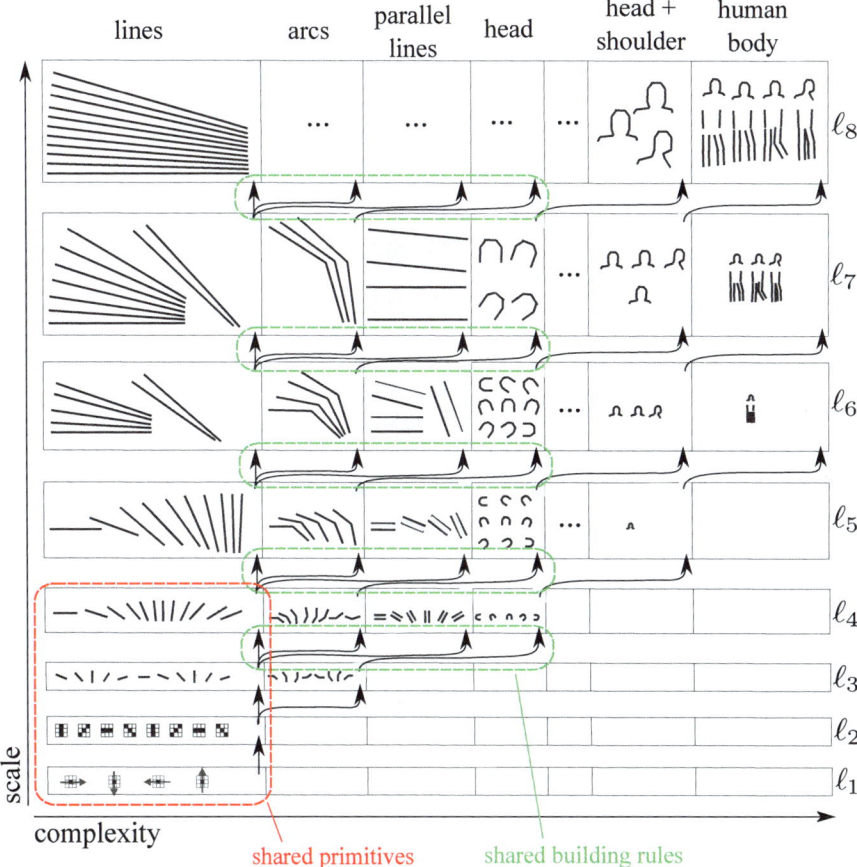

Fig. 6.5 Primitives of the learned hierarchy: Level 0 (bottom) just contains four basic gradient directions. They are combined on the higher levels to more and more complex visual primitives. The shaded rectangles (bottom) represent the training images each associated with a configuration of the human body.

Table 6.1 Complete average execution time per one training image for different scalings.

sharing	yes	yes	yes	yes	yes	yes	no
scale	1.0	1.1	1.2	1.3	1.4	1.5	1.0
time (s)	0.00891	0.00331	0.00124	0.00068	0.00045	0.00033	2.12378

shown in Fig. 6.4). We align the feature set of the model to those of the evaluation example to get the ground truth position of the whole human body. The ground truth positions of the body parts were determined by using the hierarchy in a top-down manner. Fig. 6.6 (left) shows the ROC curves for different scaling factors. Interestingly, higher scaling factors increase the detection performance, while

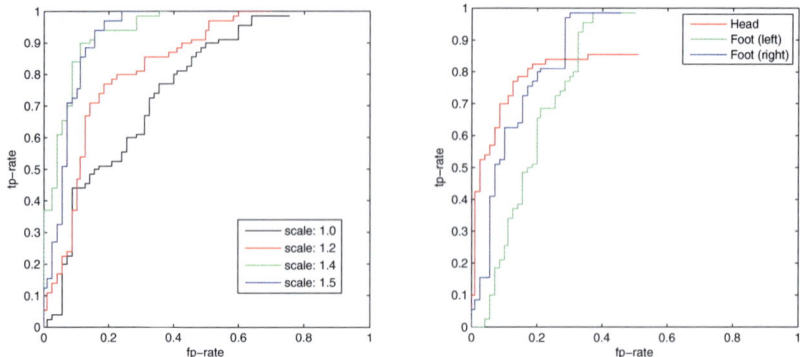

Fig. 6.6 ROCs for different scalings (left) and for the body parts (right) referring to our gait database

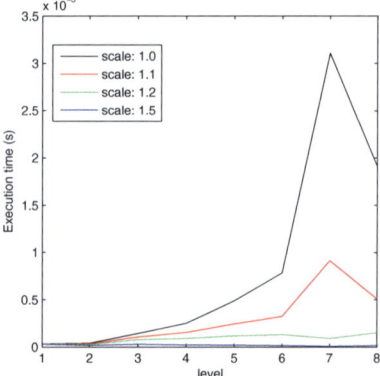

Fig. 6.7 Experimental results showing the influence of the scaling and the level on the execution time

simultaneously the execution time is reduced. The execution times can be seen in Fig. 6.7 for different scales and levels. Due to the large number of nodes the execution time increase exponentially with level. However, scaling allows to reduce the computational effort and thus the execution time becomes approximately constant for each level. The ROC curves for the body parts are shown in Fig. 6.6 (right). The detection of the head performs quite well due to the good brightness contrast of the upper body against the background. About 40% of our database contain patients with walking frames. This produces ambiguities and reduce the detection performance of the feet as can be seen in Fig. 6.6 (right). Sharing primitives among the different hierarchical graphs allows us to speed up the average execution time from 2.1s to 8.9ms per evaluation image. Using the scaling we are even able to achieve an execution time of 0.3ms for a scaling factor of 1.5 (see Tab. 6.1). The execution

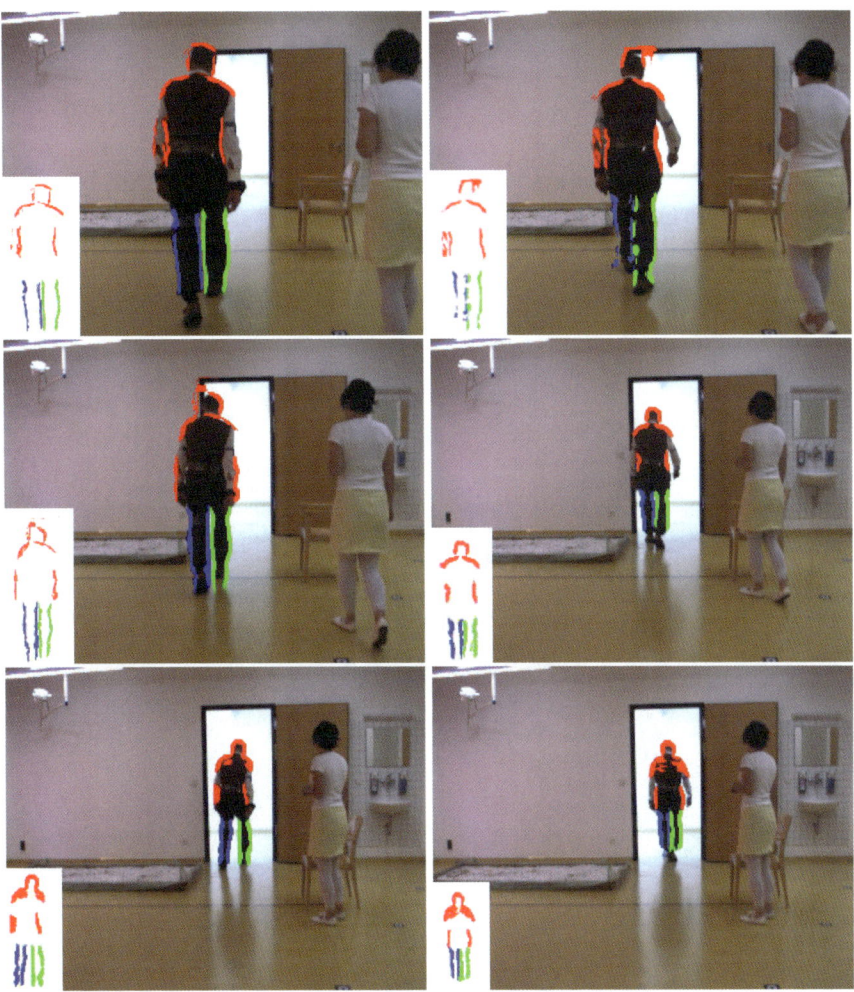

Fig. 6.8 Results for our gait database: The most likely configurations are highlighted in red (head), blue (left leg), and green (right leg).

times refer to a standard PC with 2.4GHz and 2GB RAM. The main reason for the reduced accuracy of the feet is the lower contrast of them against the background. We demonstrate the proposed hierarchical framework also using the HumanEva-I dataset. Results can be seen in 6.12.

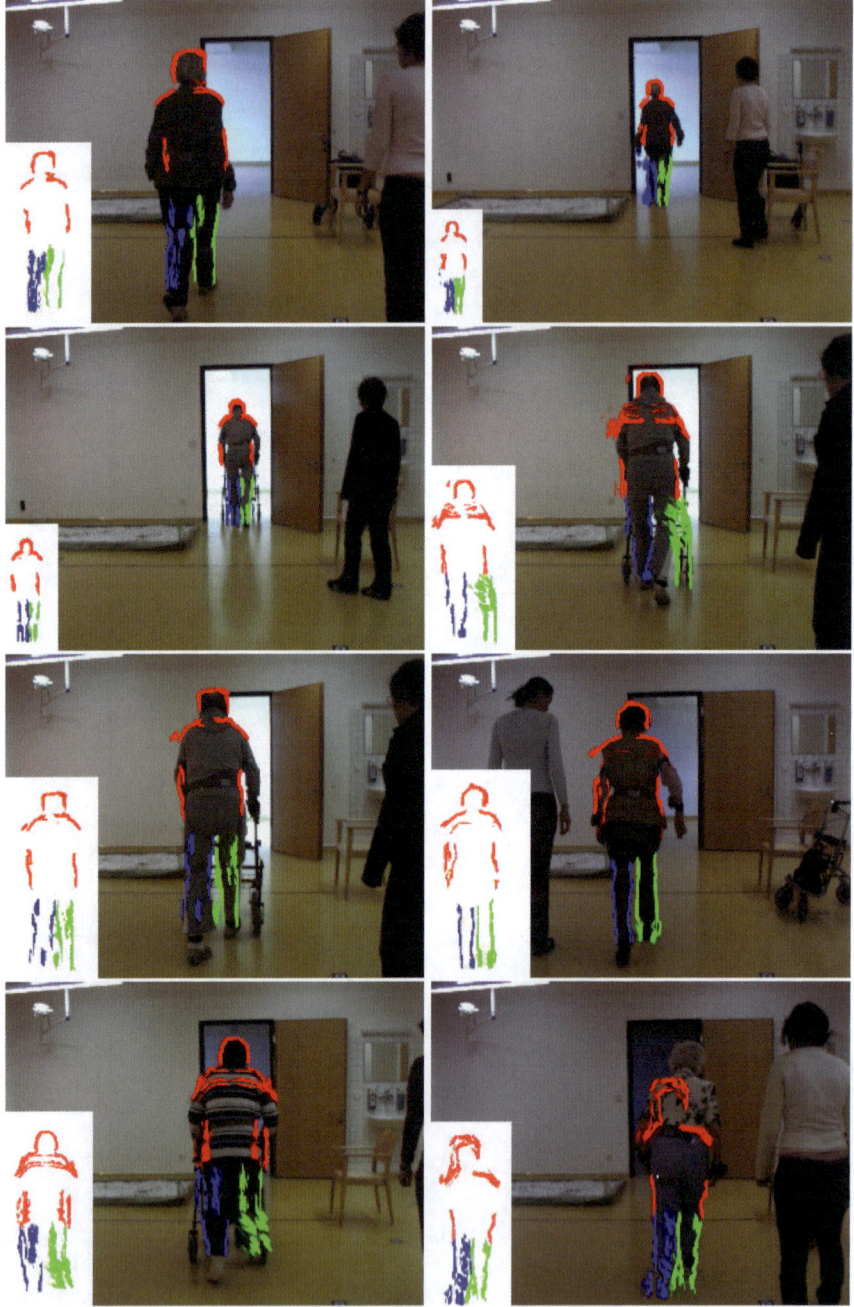

Fig. 6.9 Results for our gait database: The most likely configurations are highlighted in red (head), blue (left leg), and green (right leg).

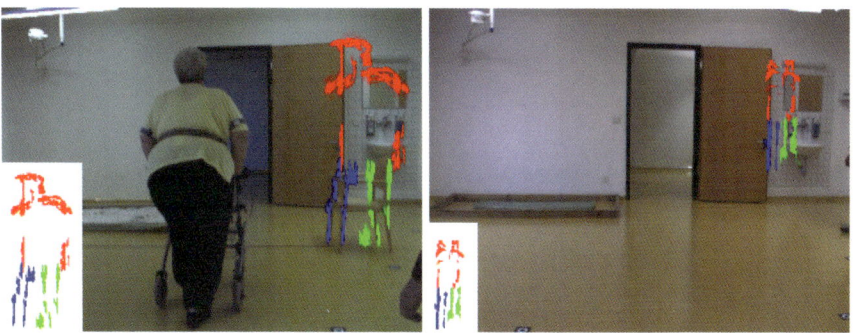

Fig. 6.10 Results for our gait database: False positives.

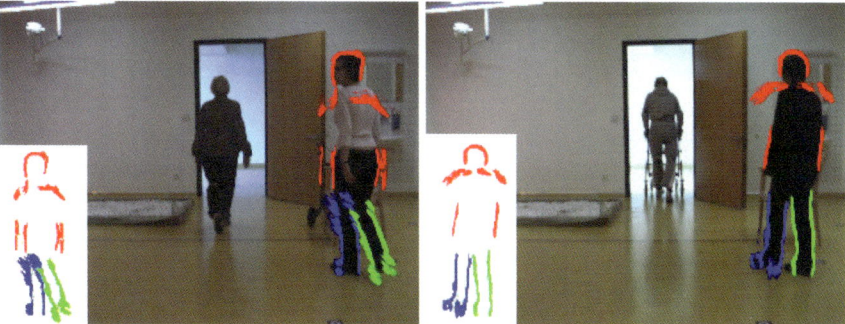

Fig. 6.11 Results for our gait database: therapist.

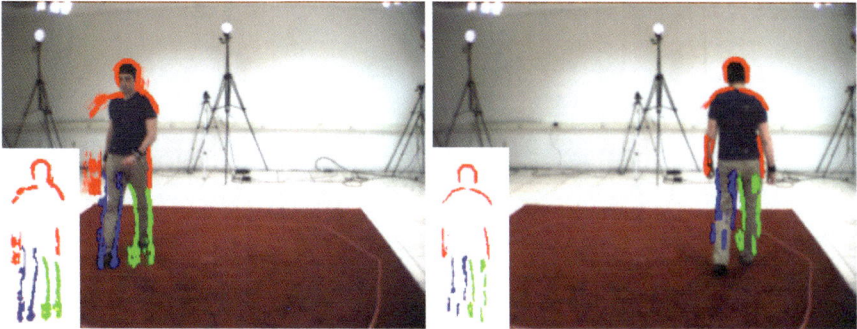

Fig. 6.12 Results for the HumanEva-I dataset

Chapter 7
Human Behavior Analysis

Due to the wide range of applications human action recognition and its representation is a popular research topic. The aim of action recognition is to automatically identify the action of a person based on some kind of sensor data. In this monograph, we focus on vision sensors that provide a stream of images over time. Detection of human actions or activities based on a video stream is challenging mainly due to the following two processing steps. First, the person has to be detected within the images and its pose has to be estimated, which is very complex due to the many degrees of freedom of the human body (see Chap. 6). In the context of action recognition the pose estimation step is often replaced by the calculation of a motion descriptor. Second, the sequent pose or motion descriptors have to be set into temporal relations in order to identify the underlying action. Unfortunately, these relations are very complex and difficult in real world scenes.

7.1 Related Work

In the following we will briefly summarize related works. A more general overview of human motion capture, analysis and its representation can be found in [149, 150, 183, 1]. A very important property of actions is its hierarchical nature. We use the action hierarchy notation as proposed in [150]: action/motor primitives, actions, and activities. Action primitives or motor primitives are atomic entities that are used to build actions. Actions are, in turn, combined to activities. Behavior is another term that is often used in the context of activity recognition. In this monograph we will use the term to describe a composition of activities. Behavior patterns are therefore a high-level representation of complex sequences of low-level actions.

7.1.1 Low-Level Action Primitives

Most of the following approaches represent actions at a low-level. Polana and Nelson [182] used spatiotemporal templates of motion features to recognize action primitives like walking and climbing. Efros et al. [46] recognized simple actions

© Springer International Publishing Switzerland 2015   135
J. Spehr, *On Hierarchical Models for Visual Recognition & Learning of Objects, Scenes, & Activities,*
Studies in Systems, Decision and Control 11, DOI: 10.1007/978-3-319-11325-8_7

like running and walking in low quality video streams, where the person is only 30 pixels tall. They used a simple normalized-correlation based tracker to get a figure-centric sequence and calculated features based on blurred optical flow. Action is classified by matching these features with a database. Ali and Shah [4] used optical flow as well. They derived kinematic features like divergence, vorticity and symmetry from the optical flow and used them to specify spatiotemporal patterns. Natarajan and Nevatia [163] extended the work of Efros et al. [46]. They explored a set of possible windows at each time step and considered large scale differences in order to improve the robustness. Different from [46], where the optical flow was directly used as feature representation, Riemenschneider et al. [191] proposed to get binary stable optical flow volumes using a maximally stable volumes detector. These sets of binary optical flow volumes were used as features in a 3d shape context descriptor. Another low-level representation of actions are *motion history images* [18], where each pixel describes the motions occurred at that point during the previous time steps. Since actions are represented by images simple template matching can be used to recognize actions. Laptev and Lindeberg [118] defined *space-time interest points* as local structures in space-time, where the image values have significant local variations in both space and time. The interest points are detected using an extension of the Harris interest point detector. The primitive descriptor characterizes the spatio-temporal neighborhood of the primitive and is built from normalized spatio-temporal Gaussian derivatives. Laptev and Lindeberg called the primitive descriptor local jet. A compact and view-invariant representation of action primitives was introduced by Rao et al. [187]. The primitives are defined in units called *dynamic instants*. Dynamic instants were computed from discontinuities of 2d trajectories of body parts like e.g. the hand. The primitives were described using a parameter 'sign', which represents the change of the motion direction at the instant. Furthermore the time period between two dynamic instants was used. Similar to dynamic instants are *key poses* that were used by Reng et al. [190]. The key poses are found based on the curvature and covariance of the normalized trajectories. Lu and Ferrier [136] used simple *linear dynamic models* to define action primitives. These primitives were automatically detected using a two-threshold, multidimensional segmentation algorithm applied to complex motions. The primitive descriptor consists of two matrices which describe the deterministic and the stochastic part of the motion. Another feature detector was proposed by Dollar et al. [42]. They detected spatiotemporal interest points by applying a quadrature pair of 1D Gabor filters to the temporal dimension of each image point and searched for local maxima of the response function. The detected interest points were described by *cuboid* descriptors, which contain the spatiotemporal neighborhood.

7.1.2 Action Recognition

In more natural scenes the short action primitives are combined to complex activities. Generally, the direct estimate of the posterior probabilities allows discriminative approaches (e.g. conditional random fields [217] or support vector machines [207]) to achieve better classification performance than generative ones. However, in this work we focus on generative approaches since they do not require simultaneous consideration of all data from all action classes and allow flexible hierarchical modeling.

High-level actions are often modeled as stochastic finite state machines like Hidden Markov Models (HMM) [184, 257]. They divide actions into discrete states, which cannot be directly observed. Instead, an emission distribution determines which output is generated by the hidden states and furthermore, a transition distribution controls the state transitions between consecutive time steps. Other variants of HMM include coupled HMM [22], layered HMM [174], multi-observation HMM [255] or hierarchical HMM [67]. Another powerful generative model is the latent Dirichlet allocation (LDA) [17]. LDA is a three-level hierarchical Bayesian model that uses Dirichlet priors to explain sets of observations by unobserved groups. LDA and its extensions have already been successfully applied to action representation e.g. [167, 243, 93]. Robertson and Reid [194] enhanced the work of Efros et al. [46] with a hierarchical combination of belief networks and HMMs. As in [46] the images are first aligned using a mean shift tracker to get a figure-centric sequence. Then, features such as trajectory information (position and velocity) and local motion descriptors are calculated. Finally, actions are classified based on a set of predefined HMMs.

Most similar to our idea of the hierarchical representation of activities are the following approaches. The work of Niebles and Fei-Fei [166] was inspired by the hierarchical model proposed by Bouchard and Triggs [19]. Their model is a constellation of bag-of-features and combines spatial and spatiotemporal features. Similar, Ke et al. [109] extended the idea of pictorial structures to video volumes. They modeled the geometric configuration of the parts and matched them in space and time. The matching of parts is mainly introduced to deal with over-segmented regions. Ryoo and Aggarwal [199] described high-level activities based on its sub-events and their temporal, spatial, and logical relationship. The activities are modeled using a context-free grammar based representation, and are classified into the categories: atomic action, composite action, and interaction.

More recently, contextual information was integrated into the action recognition process in order to improve the recognition performance. Contextual information is not only desirable but also necessary for applications like abnormal behavior patterns detection [255, 248]. A lying person is e.g. normal in the spatial context of a 'bed' or the temporal context of a 'night' but abnormal in the spatial context of a 'kitchen'. Li and Fei-Fei [55] classified events in static images by integrating scene and object categorizations using a generative graphical model. In [221] we proposed a new Gaussian feature map representation of behaviors, where the spatial and temporal context is divided into grid cells. Each grid cell models the action primitives with a Gaussian distribution of features like height, duration, magnitude and orientation of the velocity. Contextual information can also improve the action as well as object recognition performance when humans interact with objects. Moore et al. [151] combined HMM with object context to perform action recognition and object classification. Peursum et al. [180] used the spatial context and activity recognition for indirect object recognition in indoor wide-angle views. A similar approach was proposed by Gupta and Davis [84]. They described a Bayesian approach to the joint recognition of objects and actions based on shape and motion. Ryoo and Aggarwal [200] integrated human-object interactions into the hierarchical framework proposed in [199]. In order to facilitate the object or context recognition process, often additional information sources are used. Wu et al. [254] proposed an 'object-use' based activity model built of dynamic Bayesian networks, in which the nodes represent activities, objects, and sensor information. In their work they used video information as well as RFID tags to identify the objects. Marszalek et al. [143] used scene captions from movie scripts, which usually provide information on location and day time, to define contextual information. Shechtman and Irani [213] introduced a similarity measure for space-time behavior-based correlation. This measure correlates small space-time video segments against the video sequences and can be used to detect motion patterns like ballet movements or pool dives. Other related works include the coarse-to-fine hierarchy proposed by Jiang et al. [103]. Their shape-motion prototype tree represents actions and is used during testing for an efficiently search. Zhang and Tao [264] used the slow feature analysis [250] for action recognition. Action sequences are represented by accumulated squared derivative features which are feature vectors where the squared first order temporal derivatives are accumulated over all transformed cuboids. This representation is also biologically inspired, since the temporal slowness principle seems to be a general learning principle in visual perception as found in neuroscience experiments. Yao et al. [258] classified and localized human actions in video using a Hough transform like voting scheme. The Hough transform is combined with random trees for efficient voting in the spatio-temporal-action Hough space.

As can be seen, previous activity recognition approaches represent a wide range of creative combinations of well-known techniques from the field of computer vision as well as machine learning. Although the results confirm that the chosen combinations are appropriate to the particular dataset they are applied to, they are generally not trying to model the hierarchical nature of activities including the wide range of different motion pattern complexities. On the other hand, hierarchical approaches like hierarchical HMM [67] or hierarchical Dirichlet processes [243] are disregarding the continuous spatiotemporal dependencies among action primitives, actions and activities (hierarchical HMM use finite discrete-states and hierarchical Dirichlet processes use the bag-of-words assumption). Furthermore, most of the previous approaches solve action recognition in multiple stages. The person has first to be detected and tracked to get a figure-centric sequence allowing a position invariant description of the action [46, 194, 163]. Thus, the action recognition result can only be as good as the tracking result. Moreover, most of the previous approaches just use finite discrete-state machines to describe actions [257, 22, 174, 255, 67]. Unfortunately, actions are associated with continuous parameters like position, direction, duration or other motion specific descriptions, which cannot be modeled in discrete-state machines. Another drawback of finite state machines is that they just allow the description of one action at a time. Although other hierarchical representations like the ones of [199, 200] are basically able to represent sequential and concurrent

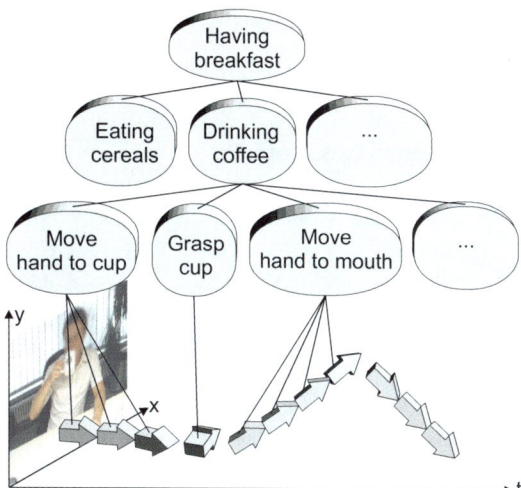

Fig. 7.1 Hierarchical nature of human actions: This figure shows schematically a spatio-temporal hierarchical graphical model with hidden nodes and observed nodes representing motion primitives detected in space-time.

actions, they are not using an unified probabilistic framework. Their applicability in real world scenarios is therefore questionable.

We introduce a new hierarchical framework which combines the localization and the detection of actions in an unified probabilistic framework. The idea is to transfer the concept of compositional hierarchies to activity representation and thus to consider actions as hierarchies of basic action primitives in the space-time volume. Each action primitive, action and activity is defined by a spatiotemporal position and a Markov random field defines the dependencies between them.

Our contribution is threefold: First, we introduce a new action primitive detector which uses dense optical flow and image segmentation methods to find atomic motions. These features are very robust and even more importantly not that rare than other features, like e.g. space-time interest points [118]. Second, we integrate the action primitives in a compositional hierarchical representation in order to combine several primitives to more complex high-level actions. We regard action recognition as an inference problem in a hierarchical Markov random field. The inference is solved using nonparametric belief propagation and the results show that the action recognition is able to simultaneously detect and localize actions at different abstraction levels. Third, the unsupervised learning scheme described in Chap. 4 is applied to video sequences; it allows us to learn robust and efficient activity models of complex long term datasets even based on single learning instances.

7.2 Hierarchical Action Representation

Human actions in real world scenes are in general very complex due to the hierarchical nature of activities. Activities are composed of actions, and actions are, in turn, composed of action primitives. Consider e.g. a person having breakfast: The activity 'Having breakfast' is e.g. composed of the actions 'Drinking coffee', 'Eating cereals' and 'Reading the newspaper'. The action 'Drinking coffee' is, in turn, composed of the action primitives 'Approaching the cup', 'Taking the cup to the mouth', 'Drinking', 'Putting the cup on the table' and so on. Thus action primitives are short scale events, whereas activities are larger scale events (see Fig. 7.1).

In our work actions are represented as nodes in hierarchical graphical models, where the leaves correspond to action primitives and the high-level nodes represent actions and activities. Both actions and activities are not directly observable. Information about them is exclusively inferred from the action primitives.

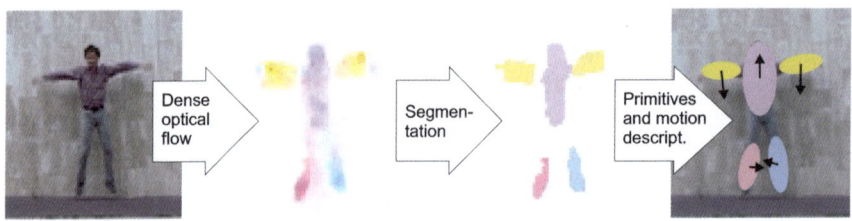

Fig. 7.2 The OFP detection steps: based on the dense optical flow result the flow field is segmented into uniform regions. For every region a position and a descriptor is calculated. The latter contains shape properties and the average motion vector.

7.2.1 Optical Flow Primitives (OFPs)

At low-level we use a new OFP detector to describe atomic motions that occur in an image sequence. Although the use of optical flow is well known in action recognition e.g.[46, 194, 163, 191, 4], our OFPs differ substantially from previous approaches. Till now, optical flow patterns were used as a whole and thus had to be figure-centric. However, we aim to split the optical flow into small motion primitives which are by themselves less meaningful. The significance of the representation is obtained by the hierarchical combination of the OFPs (see Sec. 3.5.1). The aim of the decomposition is to yield a representation that is more flexible, more robust and has also a higher structural clearness.

The OFPs are defined by the pair $f_i = (\mu_i, \delta_i)$ with $\mu_i = (x_i, y_i, t_i) \in \mathbb{R}^3$, where x_i, y_i are the position of the OFP in the image and t_i is the time step, at which the OFP occurred. Of course other dimensions could be added like e.g. the motion direction. Furthermore, each OFP has a descriptor $\delta_i \in \mathbb{R}^n$ which describes which motion occurred at position μ_i. The n dimensions of the descriptor are e.g. motion duration, motion curvature, body size or other motion/body specific properties. It is important for the descriptor to be invariant in space-time, so that specific motions will have the same descriptor independent of the image position and the time step. In the following we describe the primitive detector and the descriptor in detail.

One of the main requirements for robust action primitive detection is the invariance w.r.t problems like e.g. light changes or changing environment conditions. The detector should find the primitives in real-time and without any foreground-background segmentation. An appropriate solution for these kind of problems is the use of dense optical flow methods. Fortunately, there exist real-time implementations [262], which are supported by modern graphics processing units.

The output of dense optical flow for two images I_0 and $I_1 : (\Omega \subseteq \mathbb{R}^2) \to \mathbb{R}$ is a disparity map $u : \Omega \to \mathbb{R}^2$ that contains for each element a motion vector. Since the number of motion vectors is quite large we group them into sets containing similar motions. This segmentation problem is well known in image processing. Most of the segmentation approaches rely rather on finding groups with similar color values than groups with similar motion vectors. But the approaches can easily be adapted to our problem by replacing the color values with the 2d motion vectors. We compare two segmentation approaches: mean shift segmentation [31] and region merging [170]. In our experiments both methods perform quite well. We skip a detailed analysis of both methods since this would be beyond the scope of this monograph and it would not be much conducive to the matter of this work. However, we find the region merging approach to be much faster, and thus this method is used in the following sections.

The segmentation results in a partition of Ω in disjunctive nonempty subsets Ω_1, Ω_2, ..., Ω_P with $\bigcup_{i=1}^{P} \Omega_i = \Omega$. Each subset Ω_i is used to calculate a primitive position μ_i and a motion and shape descriptor δ_i. The position of the primitive is defined by the center of the region and the current time step. We use second order statistics of the regions to determine an appropriate shape descriptor. The descriptor contains the orientation ς_i of the region's main axes, the size in direction and perpendicular to the main axes a_i and b_i, which can be determined using the eigenvalues and eigenvectors of the corresponding covariance matrix Ψ_i. We also calculate the average optical flow vector $v_i = (\delta x_i, \delta y_i)$. A descriptor is thus defined by $\delta_i = \{\Psi_i, v_i\} = \{\varsigma_i, a_i, b_i, v_i\}$ (see Fig. 7.2). Furthermore, we define the distance between two OFPs as follows

$$d_{w_p, w_s, w_v}(f_1, f_2) = \left((w_p \Delta_p(\mu_1, \mu_2))^2 + (w_s \Delta_{skl}(\delta_1, \delta_2))^2 + (w_v \Delta_v(\delta_1, \delta_2))^2 \right)^{0.5}$$

$$(7.1)$$

where $\Delta_p(\mu_1, \mu_2) = \|\mu_1 - \mu_2\|$, $\Delta_v(\delta_1, \delta_2) = \|v_1 - v_2\|$ and Δ_{skl} is the symmetrised Kullback and Leibler divergence between zero mean Gaussians with covariance matrices Ψ_1 and Ψ_2. The factors $w_p, w_s, w_v \in [0, 1]$ weight the different terms and satisfy $w_p + w_s + w_v = 1$.

Similar to eq. 5.2 and eq. 5.3, we define the distance between two feature sets $\mathcal{F}_1 = \{f_1^1, ...f_{N_{\mathcal{F}_2}}^1\}$ and set $\mathcal{F}_2 = \{f_1^2, ..., f_{N_{\mathcal{F}_2}}^2\}$ as an extension of the modified Hausdorff distance as

$$dist(\mathcal{F}_1, \mathcal{F}_2) = \max(h_{mhd}(\mathcal{F}_1, \mathcal{F}_2), h_{mhd}(\mathcal{F}_2, \mathcal{F}_1)) + \zeta \frac{|N_{\mathcal{F}_1} - N_{\mathcal{F}_2}|}{N_{\mathcal{F}_1} + N_{\mathcal{F}_2}} \quad (7.2)$$

with

$$h_{mhd}(\mathcal{F}_1, \mathcal{F}_2) = \frac{1}{N_{\mathcal{F}_1}} \sum_{f_i \in \mathcal{F}_1} \min_{f_j \in \mathcal{F}_2} d_{w_p, w_s, w_v}(f_i, f_j) \quad (7.3)$$

where the parameter ζ additionally penalizes sets differing substantially in the feature number.

We formulate a codebook (or vocabulary) to define a finite set of OFPs. The vocabulary is learned based on the set of descriptors, which are detected and calculated in the training data. We use the k-means algorithm and the Euclidean distance metric to find clusters of similar descriptors. The cluster centers define prototypes of descriptors which represent the OFP vocabulary. After training, newly detected OFPs are assigned to the nearest element of the vocabulary using again the Euclidean distance metric. Thus, we get a low-level representation of the video based on sets of OFP positions and assigned vocabulary elements. In the following the detected OFPs will represent noisy local observations y_i that are associated to a hidden random variable x_i.

7.2.2 Hierarchical Graphical Models

The model used in this work is represented by a set of hierarchies $\mathcal{G} = \{\mathcal{G}_1, ..., \mathcal{G}_N\}$, which are specified by undirected tree-structured graphs $\mathcal{G} = (\mathcal{V}, \mathcal{E})$ as described in Sec. 3.7. The graph $\mathcal{G}$ is defined by the set of nodes $\mathcal{V}$, and a corresponding set of edges $\mathcal{E}$. As already discussed in previous chapters, the Markov random field associates each node $i \in \mathcal{V}$ with an unobserved, or hidden, random variable x_i, as well as a noisy local observation y_i. However, in this work we consider hierarchical graphical models, where just the low-level nodes are associated with a noisy local observation. All other high-level nodes are unobserved and evidence is exclusively obtained via the low-level observations y_i. The hidden high-level random variable nodes represent action primitives, actions and activities. In general, actions and activities have the same form as primitives $x_i = (x_i, y_i, t_i) \in \mathbb{R}^3$, where x_i, y_i are the position of the action primitive in the image and t_i is the time step, at which the action occurred. The neighborhood of a node $i \in \mathcal{V}$ is defined as $\Upsilon(i) = \{i | (j, i) \in \mathcal{E}\}$. Each edge $(i, j) \in \mathcal{E}$ is associated with a spatiotemporal relation between node i and j. One main advantage of this representation is, that arbitrary dependencies between actions can be modeled in space-time, even concurrent actions.

Since we are interested in high-level activities our aim is to propagate the low-level action primitive information to high-level activity information. Thus we are interested in inferring the posterior marginal distribution of an activity conditioned on the action primitives $y = \{y_1, ..., y_N\}$. For that, we apply the inference techniques as described in Sec. 3.8. Fig. 7.3 shows basically the process of the bottom-up message passing using the 'jack' action [81]. The hierarchical representation is simplified as a four level action hierarchy. The root node at level 4 represents the whole motion pattern. It has two children nodes at level 3, one representing the movement of the legs and one representing the movement of the arms and the torso. At level 2 the movements are further decomposed into e.g. the upward and downward movements of both arms. They are at level 1 decomposed into the movements of the right and the left arm. Finally, at level 0 the movements are decomposed into sets of OFPs. A detailed step of the upward message-passing is also shown in Fig. 7.4.

7.2.3 Unsupervised Learning of Hierarchies

The learning approaches proposed in Sec. 4.2 can be easily applied to the learning of hierarchical representations of activities. As before, we assume that the training data is given by activity instances $\mathcal{O}_n$ but now in terms of video sequences. In the previous sections each instance was a set of 2d image features, now each instance is a set of 3d action primitives with associated positions in space-time. For learning of the activities we will use online learning (see Sec. 4.2.2) due to its adaptability in real world applications, its good performance as already found in Sec. 5.3.1.2, and in order to show how the performance changes with increasing numbers of training samples. The online learning uses the Metropolis sampler to get samples from the posterior distribution $p(\mathcal{G}, \theta_\mathcal{G}|\mathcal{O})$. For each training instance it builds a hierarchy in a top-down manner and chooses finally that sample which maximizes $p(\mathcal{G}, \theta_\mathcal{G}|\mathcal{O})$. This scheme corresponds to the one used for learning 2d object representations. In the following we therefore just highlight differences. For a further detailed description we refer to Sec. 5.2.2.

Segmentation: Again, a set of state vectors $z = \{z_\ell\}_{\ell=1}^{\ell_{top}}$ is used to guide and facilitate the decomposition. The state vector $z_i = (w_i, \mathcal{K}_i)$ associated to element $\mathcal{A}_i$ of the auxiliary feature set includes a weighting vector $w_i = (w_p, w_s, w_v)$, $w_p, w_s, w_v \in [0,1]$, and a set of mean features $\mathcal{K}_i = (s_1, ..., s_k)$. The weighting vector scales the position, the shape and the motion components of the features. Each feature $f_j \in \mathcal{A}_i$ is assigned to the nearest mean feature $\hat{k}_j = $ argmin$_{k=1,...,n}(w_p^2 \Delta_p^2(s_j, s_k) + w_s^2 \Delta_{skl}^2(s_j, s_k) + w_v^2 \Delta_v^2(s_j, s_k))$.

Proposal Function: The proposal functions $q(w^*|w^{(\tau)})$, $q(k^*|k^{(\tau)})$ and $q(s^*|s^{(\tau)})$ are mainly identical with those used in Sec. 5.2.2. Just the weighting vector w_i as described before and the dimensionality of the mean features were changed.

The sampling starts with the root nodes of the new hierarchies, which represent the training instances. They are put into the auxiliary feature sets at the highest level and are recursively decomposed into smaller feature sets. As before, we omit all samples, which are generated in a burn-in period of 20 samples.

level 4

level 3

level 2

level 1

level 0

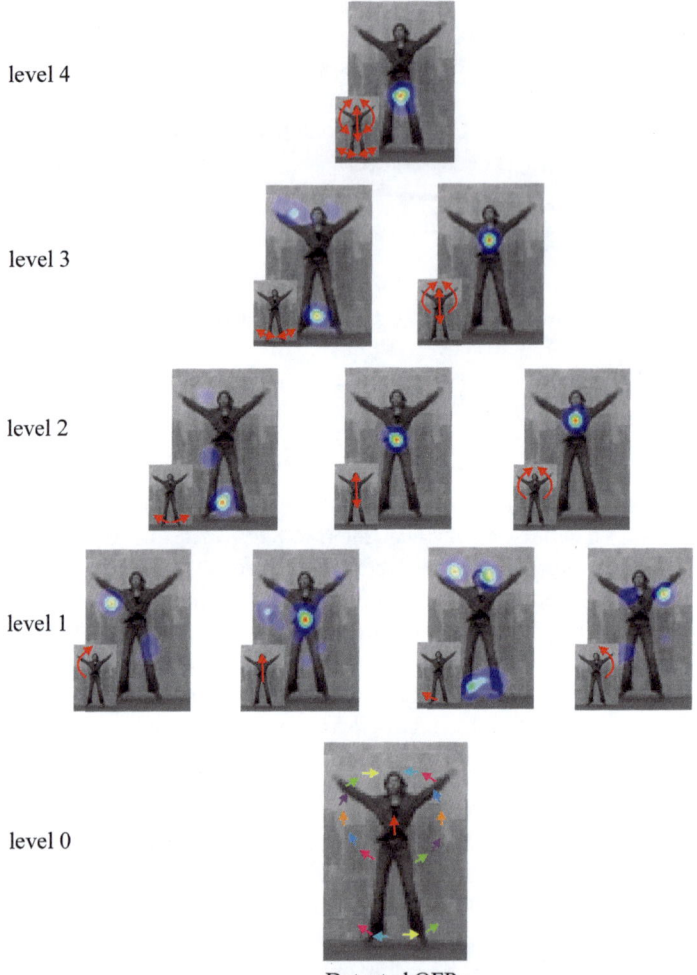

Detected OFPs

Fig. 7.3 Upward message-passing using the hierarchical model of the jack action: The detected OFPs (bottom) are used to infer the positions of action nodes on level 1 (not all nodes are shown). The beliefs of the action nodes are highlighted by the pseudo-colored PDF, in the small images the detected actions are indicated by the red arrows. The message passing is continued for the next levels (from bottom to top). The final belief of the root node shows that the hierarchical framework allows a precise localization of the action (top).

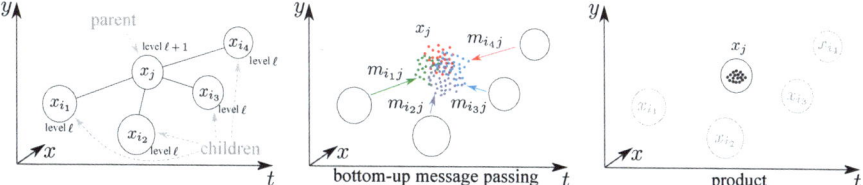

Fig. 7.4 Detail of the upward message-passing. Since the evidence is exclusively inferred from the low-level OFPs, the positions of the nodes x_i will be estimated first (left). They will send messages m_{ij} to the next higher level node x_j containing a rough estimate of the position (center). Although the messages support just an uncertain estimate, the product of the messages allows an accurate localization of the action (right).

7.3 Results

In this section we demonstrate the application of the proposed hierarchical graphical models using different datasets. We use a dataset containing hand movements. The aim of this dataset is to show, how the hierarchical representation can model the sequences of actions and thus overcome problems of bag-of-words based representations (e.g. LDA). Furthermore, we use a long term dataset containing activities of daily living where supervised learning of activities is intractable due to the large amount of video data and the large number of different activities. Here, we will show how the proposed unsupervised learning can automatically extract motion patterns at different hierarchy levels.

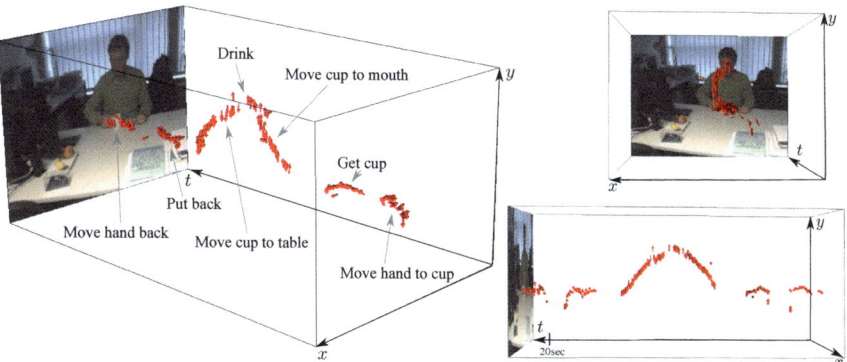

Fig. 7.5 Visualization of the activity 'Drinking coffee'. The figure shows the detected OFPs as red arrows in the x,y,t-space.

7.3.1 Hand Motions as Activities

We built up a dataset of a person performing typical hand motions captured from a frontal camera. These hand motions include activities like 'Eating fruits', 'Drinking coffee' or 'Food preparation'. The length of the sequences is typically 5-20sec. They are performed with different speeds of execution, different orders and different durations between the low-level actions. The activities are furthermore built up of similar low-level actions. Thus, the activities 'Eating fruits' and 'Drinking coffee' have for example an action 'Hand moving to the mouth and back' in common. These activities can therefore not easily be distinguished using their low-level representation. Instead, a hierarchical structure, which captures the spatiotemporal dependencies between the actions, has to be used. Here, especially the role of context comes into play. Using again the example from above, the 'Hand moving to the mouth and back' might be the same for the activities 'Eating fruits' and 'Drinking coffee'. However, the context in which it occurs is different, i.e. we will grasp the cup before drinking coffee and grasp the fruits before eating fruits. Since the hierarchical representation models also the context in terms of information provided by the parent node, it is able to easily deal with this difficult problem.

Since the corresponding hierarchies are complex 3d structures the visualization is difficult. We chose the activity 'Drinking coffee' as one example where the hierarchical structure can be obviously seen. Fig. 7.5 shows the detected OFPs in the x,y,t-space. In order to visualize the 3d structure more clearly a diagonal, front (x,y) and side (y,t) view are shown. The OFPs are indicated by red arrows, which also show the motion direction. The different sections of the activities represent simple actions like 'Move cup to mouth'; the corresponding labels are manually assigned to the OFPs.

We apply the unsupervised online learning approach as discussed in Sec. 7.2.3 to the sequences. For each activity we have 9 instances. In order to show the influence when learning from one or more exemplars we perform the learning sequentially. Thus, learning based on one training instance means that for each activity class just one training instance was given. Fig. 7.6 shows the learned hierarchies for two activities. In Fig. 7.6(a) the unsupervised hierarchy for the 'Drinking coffee' activity is shown. The high-level nodes are colored according to their hierarchy level and are manually translated in y-direction to make the hierarchical structure more obvious. The node level is determined during learning according to eq. 3.33. The root node (blue) of the hierarchy representing the whole activity is first decomposed into two parts, and then into five parts. These parts correspond approximately to those actions manually assigned in Fig. 7.5. Although the hierarchy was learned unsupervisely, the result gives a reasonable decomposition similar to manually assigned ones. At lower-levels actions represent simple linear movements like grasping or moving. Please note, that the number of children nodes is variable and that also the children do not have to lay on the next lower hierarchy level. This guarantees an efficient sharing structure. Although Fig. 7.6(b) looks quite similar it shows the activity 'Eating fruits'. It has a similar action of a hand moving to the head and back in the middle of the activity. However, the start and end differ concerning the direction of the movement.

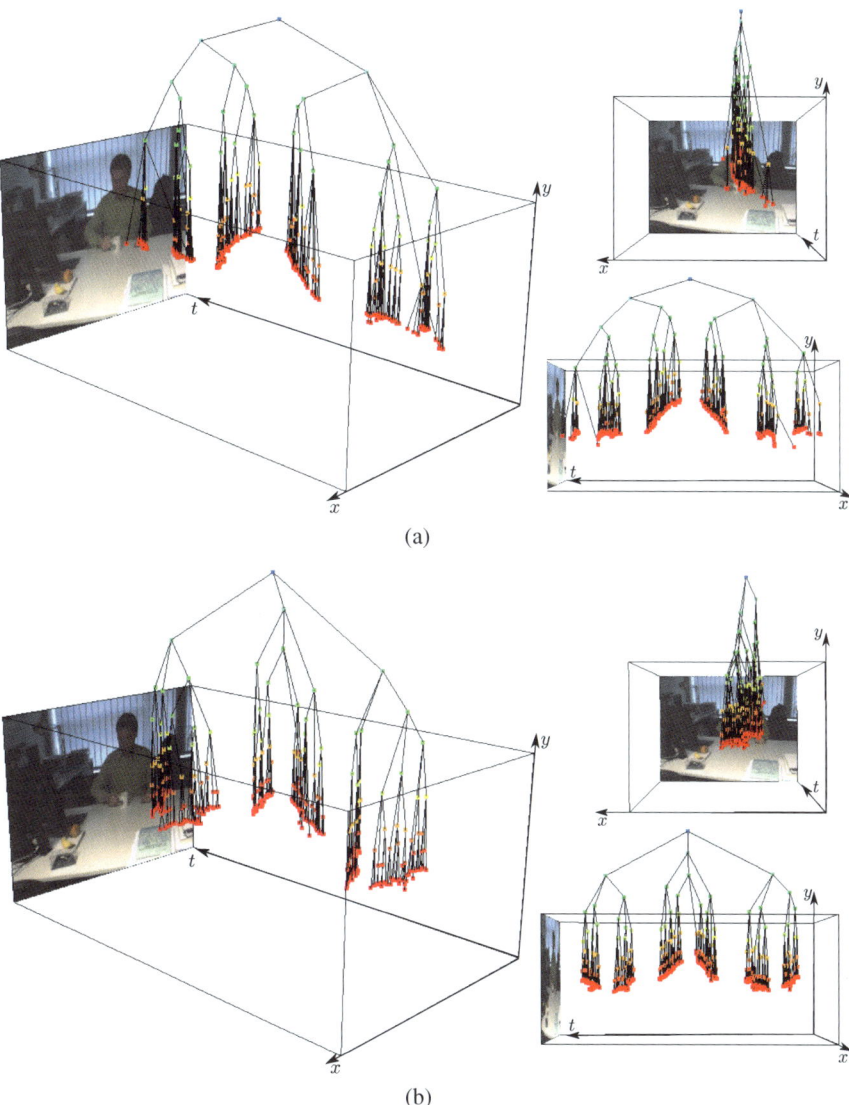

Fig. 7.6 Two examples of unsupervised learned hierarchies: (a) 'Drinking coffee' and (b) 'Eating fruits'. The high-level nodes are manually translated in y-direction to make the hierarchical structure more obvious. The red nodes represent the OFPs while the other nodes represent the hidden action and activity nodes with hierarchy level indicated by an individual color.

In order to demonstrate the influence of different sharing degrees we investigate several learning runs while varying the sharing properties. We perform the evaluation with three different sharing thresholds τ_e as in eq. 5.8

$$\tau_e(\ell) = \alpha_{\tau_e}(\beta_{\tau_e})^\ell \qquad\qquad (7.4)$$

with $\alpha_{\tau_e} = \{0.45 - 2.4\}$ and $\beta_{\tau_e} = 1.1$.

The dataset is divided into one learning and one evaluation dataset. In order to demonstrate the influence of sequential learning, we repeat the evaluation for different numbers of training samples. Each training sample contains just one instance for each activity ('Drinking' and 'Eating'). The learning from single training instances allows us to demonstrate the generalization properties of the hierarchical structure. This is especially important in order to show the influence of sharing on the generalization. The evaluation dataset contains each 72 positive and negative samples. We evaluate the performance using the F-score which combines both the precision and the recall: $F = 2 \cdot (precision \cdot recall)/(precision + recall)$. We represent each hierarchy level by a histogram containing the number of detected primitives and actions. Level 16 contains the root nodes of the activities and the histogram is therefore relying at level 16 directly on a representation of the whole activity, while the histograms at lower levels are based on low-level actions and action primitives. A nearest neighbor classifier is used to assign the histogram to an activity category, which is given by the training instances. Using this histogram based representation of activities at different levels allows us to demonstrate the advantage of our hierarchical representation against bag-of-words based representations, which just use the features of level 1.

The average F-score is shown in Fig. 7.7 for $\alpha_{\tau_e} = 0.45$ (Fig. 7.7(a)), $\alpha_{\tau_e} = 1.2$ (Fig. 7.7(b)) and $\alpha_{\tau_e} = 2.4$ (Fig. 7.7(c)). Each bar diagram plots the F-score over the number of training instances and the hierarchy level. Unsurprisingly, the classification based on level 1 features performs very poorly in all three sharing setups. Even for a larger number of training instances the accuracy does not increase. As can be seen, the accuracy increases with the hierarchy level as well as the number of training samples. The highest F-score is reached at level 16. The root nodes at level 16 model the spatiotemporal order of the actions and are therefore able to distinguish between different activities, whose low-level representation is similar. Therefore, the recognition accuracy for level 1-2 is about 0.5, for level 3-16 it is greater than 0.5. Comparing the different sharing parameters shows that the F-score at level 3-12 is slightly decreased for a high degree of sharing (Fig. 7.7(c))). This is due to the sharing between different activity classes, which minimizes the amount of discriminative features. Interestingly, the F-score for one training instance at hierarchy level 16 is higher in Fig. 7.7(c) than in Fig. 7.7(a) and Fig. 7.7(b). This indicates that the high degree of sharing increases the generalization properties.

In Fig. 7.8 one can see the corresponding execution times. The plot in Fig. 7.8(a) is typical for a multi-instance setup since the execution time increase with the number of training instances. Here, the degree of sharing is low. In contrast to this the overall execution time is much lower in Fig. 7.8(b) and 7.8(c), where sharing takes

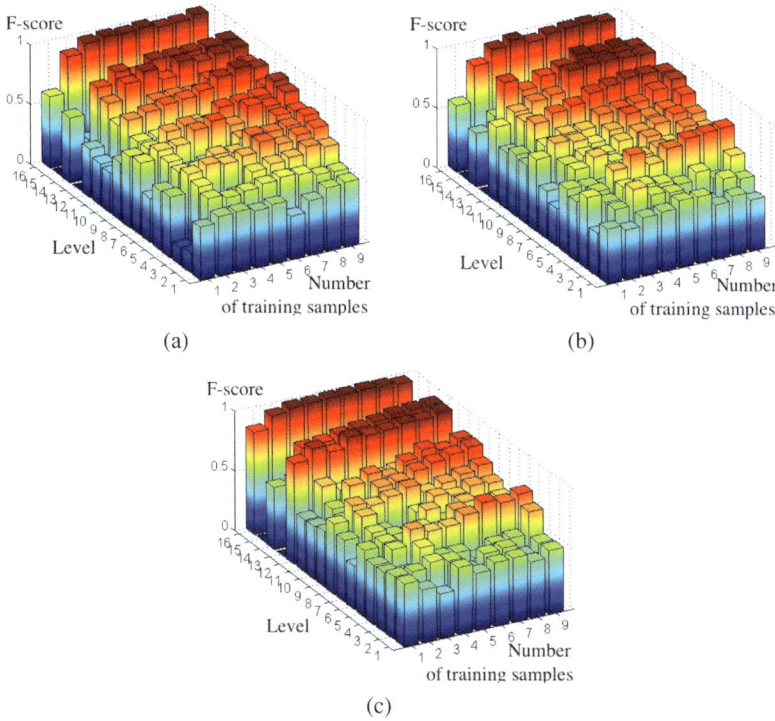

Fig. 7.7 F-score as a function of the number of training samples and hierarchy level. The underlying hierarchies are built using different sharing parameters: (a) $\alpha_{\tau_e} = 0.45$, (b) $\alpha_{\tau_e} = 1.2$, and (c) $\alpha_{\tau_e} = 2.4$.

effect. The time is just slightly increasing for more training instances. In Fig. 7.8(c) it seems even to be constant for the low-level action primitives. This relation becomes also clear in Fig. 7.9, where the number of primitives per level is shown. The plot in Fig. 7.9(a) is quite similar to the one showing the corresponding execution time (Fig. 7.8(a)). This shows again, that the execution time significantly depends on the number of primitives. The more the degree of sharing increases the more stays the number of primitives constant (Fig. 7.9(b) and Fig. 7.9(c)). Please note that for $\alpha_{\tau_e} = 2.4$ the maximal number of primitives per level reaches 40 while for $\alpha_{\tau_e} = 0.45$ more than 300 primitives are needed.

An example of the recognition results for the activity 'Drinking coffee' is shown in Fig. 7.10. Different primitives of the activity are shown and arranged according to their hierarchy level. At the bottom, four low-level features are shown. They represent the OFPs detected in the image sequence. The primitives are indicated by red arrows in the left upper corner. For each primitive we divided the x,y,t-space into two 2d views: (x,y) and (y,t), and draw the PDFs in pseudo color. As can be seen, the action primitives on level 1 as well as the actions at higher levels (e.g. level 2) are just roughly localized. The many local maxima are due to ambiguities, background

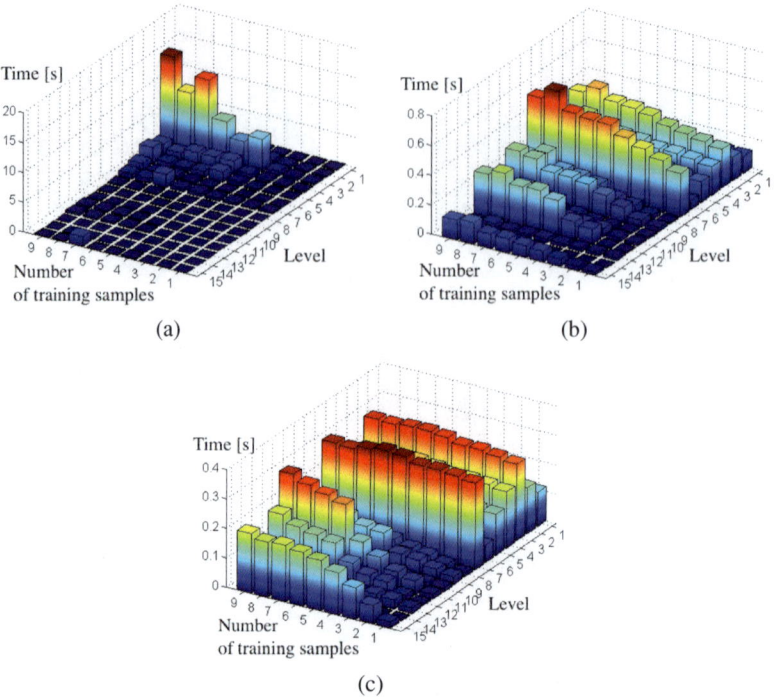

Fig. 7.8 Time as a function of the number of training samples and hierarchy level. The underlying hierarchies are built using different sharing parameters: (a) $\alpha_{\tau_e} = 0.45$, (b) $\alpha_{\tau_e} = 1.2$, and (c) $\alpha_{\tau_e} = 2.4$.

noise, shadows or segmentations errors of the optical flow. However, the activity on level 16 contains one unique maximum since the noise is iteratively reduced at the higher levels. As mentioned before, the new representation allows a simultaneous detection and localization of actions at different abstraction levels.

We also apply the hierarchical model to temporally scaled activities and compare the detection rates. The performance depends significantly on the covariance of the spatiotemporal potential functions. As described in Sec. 4.2.3, the covariance has to be predefined. Fig. 7.11 shows the detection rate for different values α_σ and β_σ. One can see, that at least $\alpha_\sigma = 3.0$ is necessary to reach a detection rate of 100% for unscaled input data. The parameter β_σ can be used to control the sensitivity to scale changes. For $\beta_\sigma = 1.0$ the curve is very peaked and reaches a detection rate of 0% even at a scaled activity of just $\pm 20\%$. For increased values of β_σ the scale invariance increases as well. For $\beta_\sigma = 1.2$ the detection rate is still greater than 0.5 at $\pm 50\%$.

In the next experiment we investigate the influence of uniformly distributed noise in terms of randomly added OFPs. Although the number of randomly added OFPs corresponds to ten times the number of the actual activity the detection rate stays constantly 100% in all of our tests. Fig. 7.12 gives an impression about the

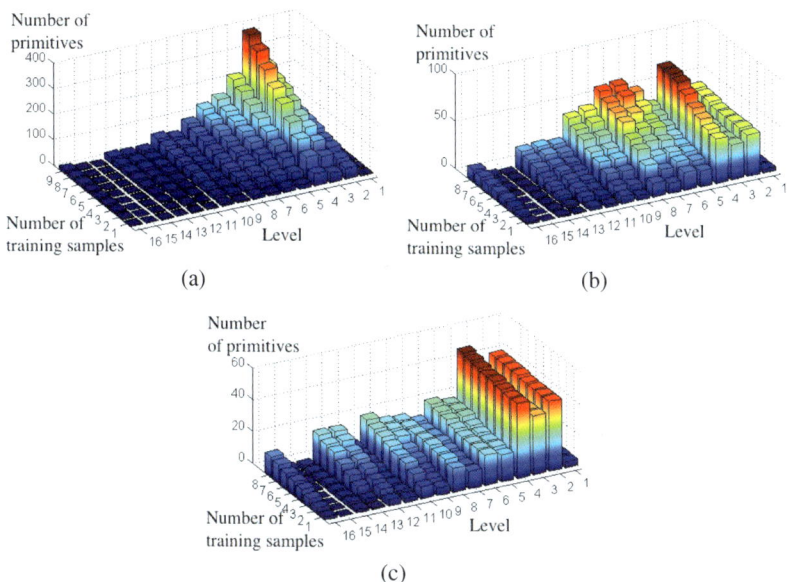

Fig. 7.9 Number of primitives as a function of the number of training samples and hierarchy level. The underlying hierarchies are built using different sharing parameters: (a) $\alpha_{\tau_e} = 0.45$, (b) $\alpha_{\tau_e} = 1.2$, and (c) $\alpha_{\tau_e} = 2.4$.

robustness against noise. The correctly detected activity is highlighted in green, while the noise is colored red.

Very important for the applicability of the representation is the robustness against missing OFPs, e.g. due to occlusions. Occlusions are very typical for activity recognition task since in most of the real world applications it is very unlikely that an action or motion primitive is continuously visible from the camera's point of view. The hierarchical representation is in itself suitable for occlusions due to the 'part'-based or action based representation. It guarantees that at least not occupied actions will be detected. Fig. 7.13(a) plots the detection rate over the missing OFPs. Three curves are shown representing different thresholds τ_ℓ. The threshold is used to stop the bottom-up propagation of unlikely hypotheses. The value can therefore be seen as a likelihood threshold. It can be used to control the sensitivity as Fig. 7.13(a) shows. Even a detection rate of 1.0 can be realized for about 70% missing OFPs. However, this increases the execution time extremely, as Fig. 7.13(b) shows.

We also added Gaussian noise to the positions of the OFPs in order to analyze the effect of different noise levels. In every dimension (x,y,t) we used a zero mean Gaussian with different standard deviations. Fig. 7.14(a) shows the corresponding influence of the detection rate. Similar to our previous experiment, the detection rate depends on the covariance of the spatiotemporal potential functions. The parameters α_σ and β_σ can be used to control the sensitivity to different noise levels. As before, larger α_σ and β_σ also increase the execution time as Fig. 7.14(b) shows.

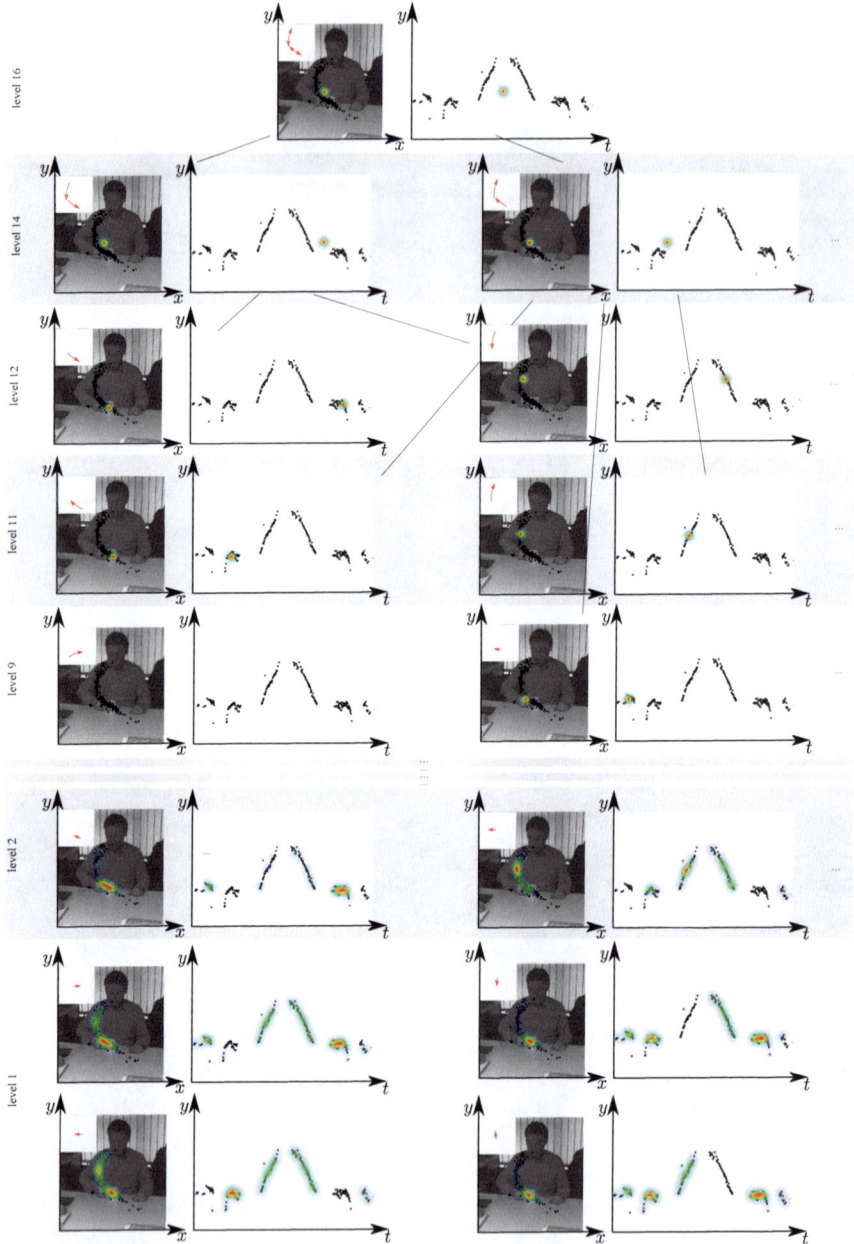

Fig. 7.10 Spatial and temporal part of the PDFs for the activity 'Drinking coffee' and for their corresponding actions: OFPs (level 1), 'Move hand to mouth' (level 11), 'Move hand to cup and then to mouth' (level 14), whole activity 'Drinking coffee' (level 16).

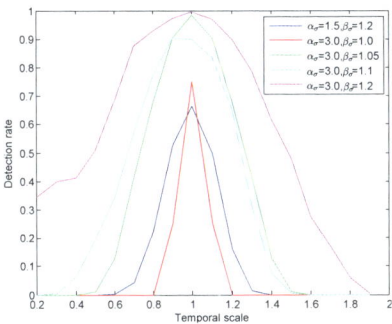

Fig. 7.11 Influence of temporally scaled input data caused by variations of the speed profile: Detection rate as a function of scaled input data.

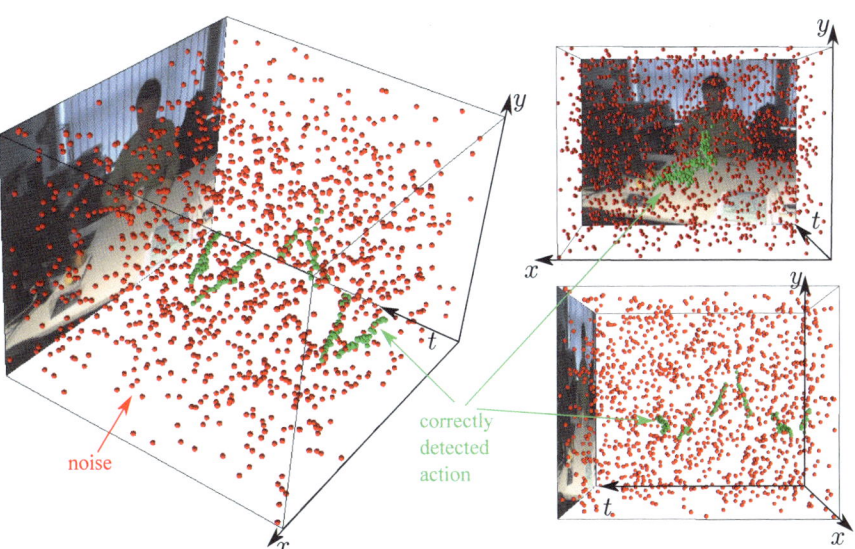

Fig. 7.12 This example shows the influence of noise. Random OFPs are added in order to distort the input data. However, the activity is correctly detected (activity is highlighted in green and the noise is colored red).

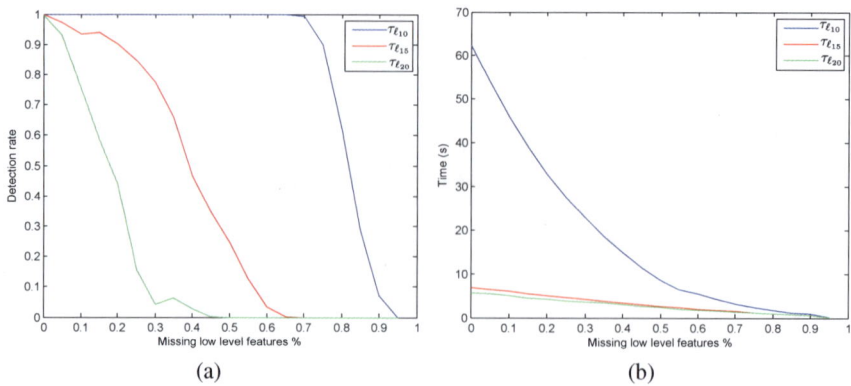

Fig. 7.13 Influence of missing OFPs: (a) Detection rate as a function of missing low-level features. (b) Corresponding execution times.

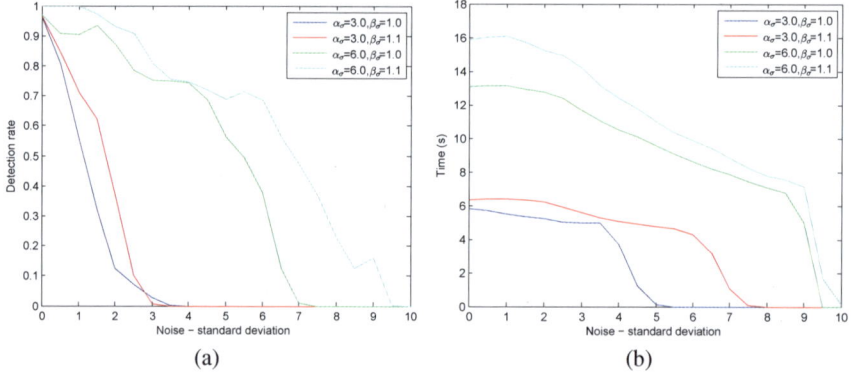

Fig. 7.14 Influence of additive Gaussian noise: (a) Detection rate depending on additive Gaussian noise. (b) Corresponding execution times.

7.3.2 Long Term Activities of Daily Living in a Home Environment

We use a long term dataset containing activities of daily living in order to demonstrate the benefits of our new unsupervised learning scheme. Due to the large amount of video data, manually labeling of the data is intractable and thus supervised learning of activities cannot be applied. We chose an one-person household of an elderly person and captured video sequence over 14 month. A fisheye camera captured the whole room with one camera image and was mounted top-down at the ceiling in the kitchen. The long term dataset allows us to extract and learn short term motion patterns as well as long term behavior patterns. Fig. 7.15 shows an example of the distorted fisheye camera image and the observed OFPs during the activity 'Having Meal'. After capture the descriptions of the actions were manually assigned. In our experiments we concentrate on this activity since it is the most complex pattern we observed in our dataset. It can be hierarchically decomposed into more simple actions like food preparation or eating. Typically, the whole activity takes about 30min including a preparation, cooking, eating and cleaning phase; it is represented by about 20.000 OFPs. As can be seen in Fig. 7.15 (indicated by the OFPs between the door and the sink region), it starts with 'Entering the kitchen' and ends with 'Leaving the kitchen'.

We use the same learning framework as before and learn sequentially instances of the activity 'Having Meal'. Fig. 7.16 shows one learned hierarchy with levels indicated by colors ranging from red (level 1) to blue (level 23). One can see that the activity is divided into three parts. The first part contains 'Entering the room', 'Cooking' and the 'Food preparation', the second part 'Setting the table', 'Sitting down' and 'Starting to eat', the third part contains 'Eating', 'Clearing' and 'Leaving the room'. At about level 11 the nodes represent actions like 'Entering the kitchen' or 'Going from the table to the stove'. Some primitives are also visualized in Fig. 7.17 where the primitives including the motion direction (blue) and the shape (red) are shown. In Fig. 7.17(a) the whole activity 'Having Meal' represented by a root node at level 23 is shown. One can see, that most of the motions were captured in front of the kitchen unit. But one can also see, that reflections of the motions at the kitchen furniture were included in the hierarchical representation. Since the approach models all observed OFPs its does not distinguish between motions caused on the one hand by human movements and on the other hand by reflections or illumation changes. The 'Clearing' and 'Leaving the kitchen' actions are shown in Fig. 7.17(b) (level 15). Fig. 7.17(c) and 7.17(d) show primitives of level 11, which consists of approximately 80 OFPs. Fig.7.17(e) shows an example of an action at level 6. Similar to most of the actions at this level, the action represents a linear movement of the person. An example of level 5 is shown in Fig. 7.17(f), where the left and right movement of the person during clearing is represented.

Sequential Learning. As in the previous section, we also apply sequential learning on several training instances, where each instance represents one observed activity 'Having Meal'. In this case sequential learning is especially attractive since it could by applied in an intelligent environment, where an observation system learns new motion patterns over time. While after the installation of the vision system no explicit training sequences are available, over time more and more sequences are captured, which can be used for online unsupervised training, so that the representation of the human behavior patterns becomes more and more accurate. At every time step the system can compute how similar new observations are compared to already seen patterns. Especially in this case the number of reusable parts is very interesting since it gives information about how good the new observation can be explained based on previous data.

We investigate several sequential learning runs and plot the changes of the hierarchies in Fig. 7.18. Here, the number of primitives is shown for each hierarchy level during sequential learning. The colored bars represent the different learning iterations. As can be seen, during learning the number of low-level features (level

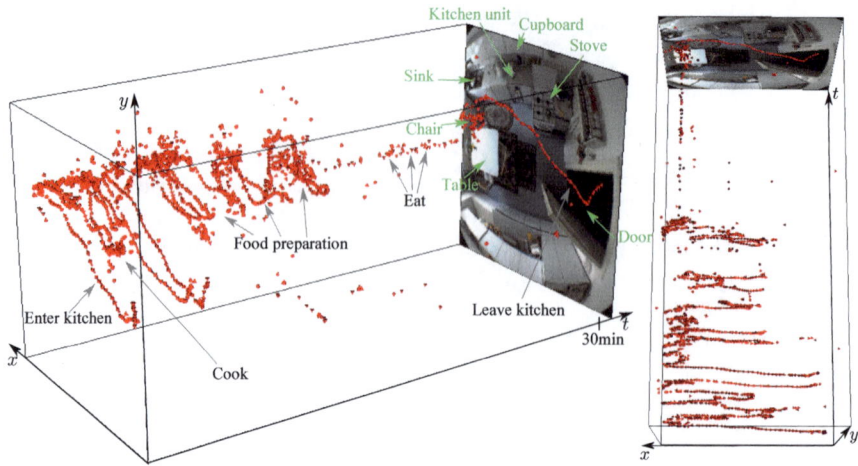

Fig. 7.15 Visualization of the long term activity 'Having Meal'. The figure shows the detected OFPs as red arrows in the x,y,t-space. The image data is captured using a fisheye camera mounted at the ceiling of the kitchen (the different regions are labeled green) and the OFPs represent trajectories of the person while going through the room and performing typical kitchen activities.

1-5) stays constant. Between level 6 and 22 the number slightly increases with the hierarchy level. At level 23 the number grows linearly with the new training steps since in each iteration one new root node was added to the hierarchy. In Fig. 7.19 we visualized some of the sequentially learned hierarchies. Blue nodes indicate reused actions that are shared with previously learned models and the green nodes represent new nodes. In our experiments primitives were shared actions up to level 13-14. The corresponding motion patterns contain each about 200 low-level primitives. Although we were able to share actions like 'Entering the room' or parts of activities like 'Food preparation', the whole activity 'Food preparation' was not shared between the different activity instances. The variations are too strong, so that reusable activities were not found. Please note, that the activities were neither performed by an actor nor according to a script. Therefore, variations were present at low as well as at high-level. The low-level variations result from different realizations of an action due to breaks or interactions with objects. For instance, the person can leave the room with closing the door or without. Although the actual activity is in both cases the same, the motion pattern represented by the observed OFPs is different, and thus modeled with different hierarchy nodes. Another example are movements that are ideally characterized by a linear motion between two positions. However, in our dataset we found that those ideal motions are quite rare. Most often the movements were combined with other actions like grasping of objects resulting in large variations of the observed motions primitives. These parallel actions are another problem during learning. Here, the OFPs detector reaches its limits since the parallel actions are often not correctly segmented into parallel motion fields. For example, when the person approaches the table and grasps simultaneously a spoon, the detector will in

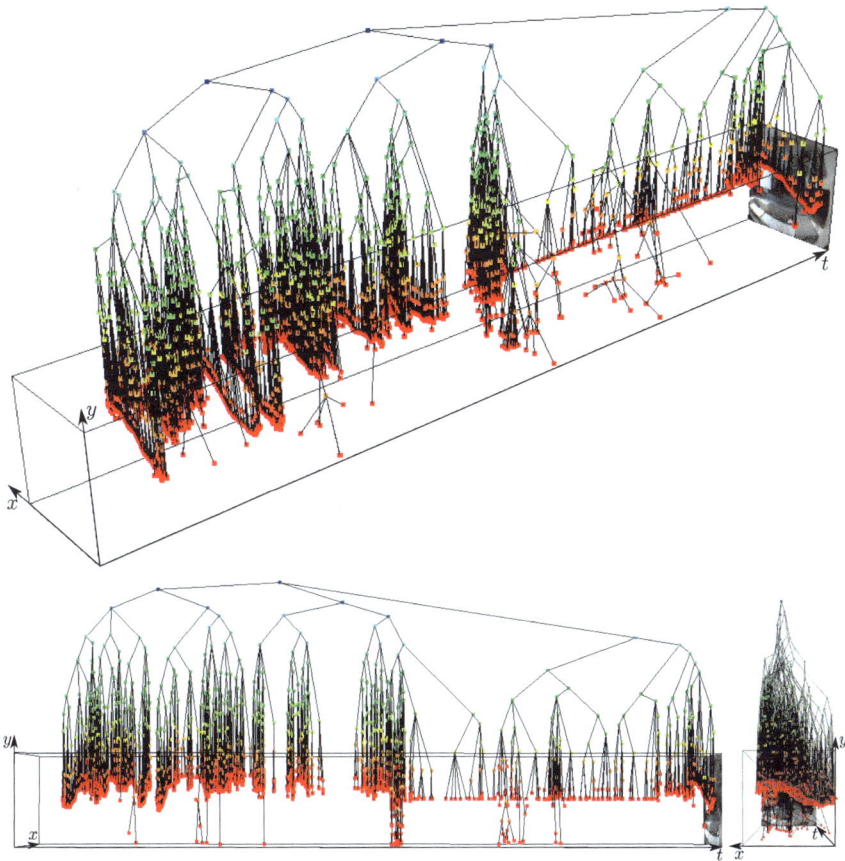

Fig. 7.16 Unsupervised learned hierarchy of the long term activity 'Having Meal'. As before, the high-level nodes are manually translated in y-direction to make the hierarchical structure more obvious. The hierarchy level is indicated by an individual color.

general detect just one OFP instead of two. Even more critical are the variations at higher levels. The main reasons for these variations are:

- The higher levels represent long term activities, so that a lower number of examples for one specific activity is available compared to actions primitives. During the activity 'Having Meal' we get one example of the activity 'Food preparation' while in general we get several examples for the action 'Going to the stove'.
- The order can be changed, and the type of actions can be combined in many different ways. During food preparation it is for example often not important in which order ingredients are cut into pieces, so that the order can change.
- Activities are often done in parallel. Some activities like e.g. 'Cooking' can be seen as background activities that do not require the person to stand beside the stove during the whole time of preparation. Thus, other activities like 'Setting the table' can be done concurrently.

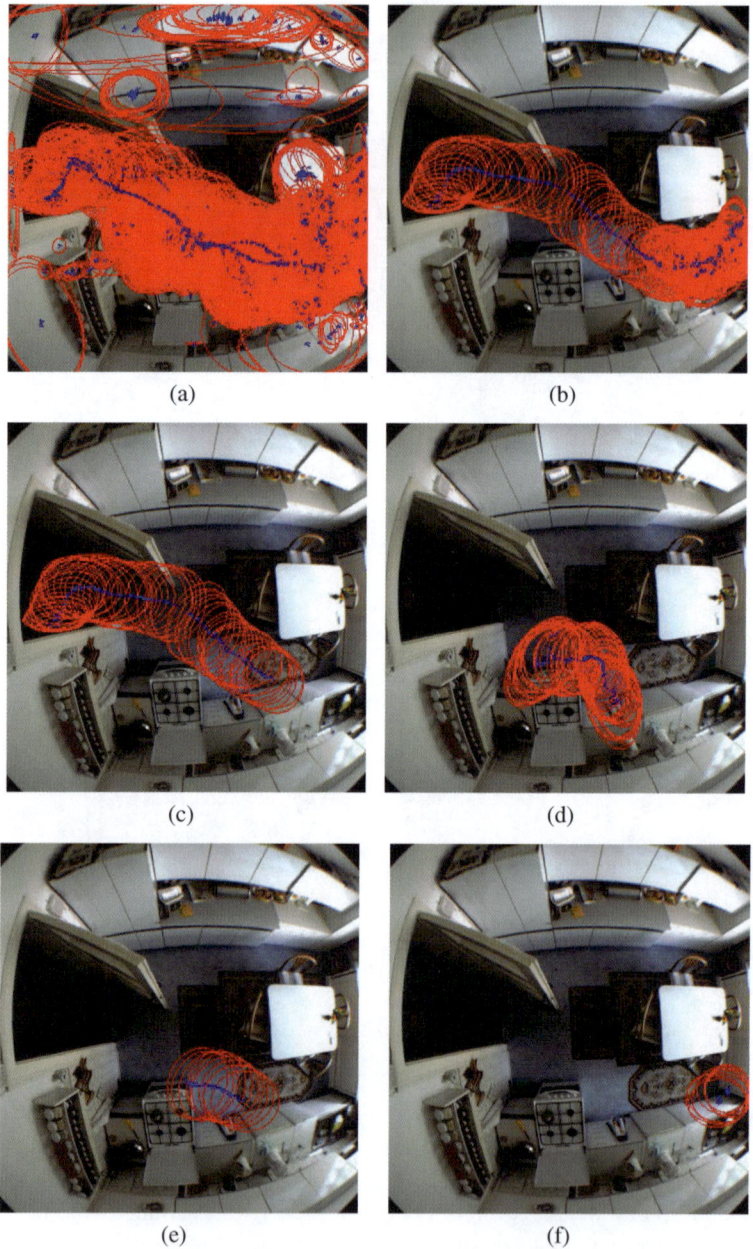

(a)   (b)

(c)   (d)

(e)   (f)

Fig. 7.17 Long term activities of daily living in a home environment: The red ellipses and the blue arrows represents OFP primitives. Each image shows one detected activity or action: (a) Whole activity 'Having meal' (level 23), (b) 'Entering the room and food preparation' (level 15), (c) 'Entering the room' (level 11), (d) 'Cooking' (level 11), (e) 'Going to the stove' (level 6), (f) 'Dish washing' (level 5).

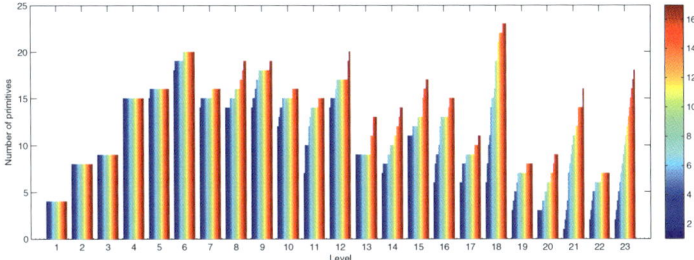

Fig. 7.18 Number of primitives for each hierarchy level during sequential learning. The iterations are indicated by color.

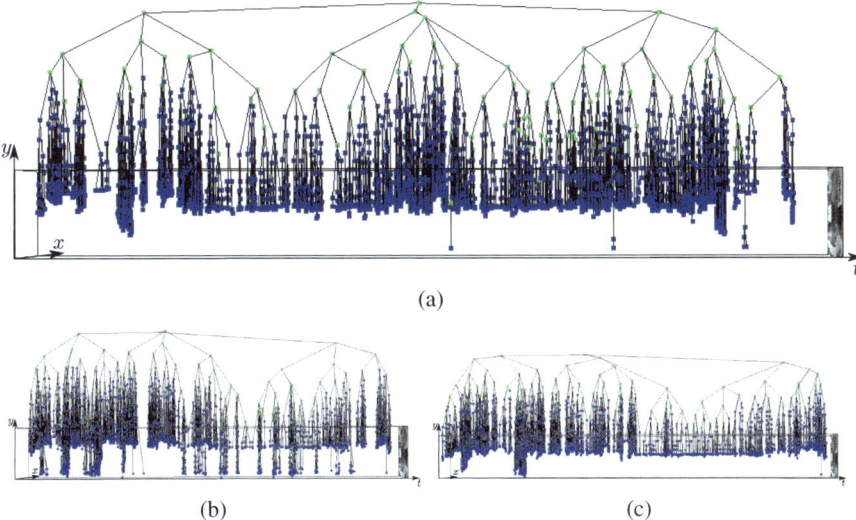

Fig. 7.19 Three examples of the sequentially learned hierarchies. Blue nodes indicate reused actions that are shared with previously learned models and the green nodes represent new nodes.

All these reasons complicate the learning of a robust representation. One solution might be the use of absolute spatial and temporal information. Up to now, we exploit just relative information into our model. The spatial as well as the temporal relations are incorporated in the edges of our hierarchy. Absolute information about spatial prior knowledge were omitted. However, absolute information might be helpful to distinguish for example the activity 'Cooking' taking place at the stove and 'Doing the dishes' taking place at the sink. This absolute information could be fed into the hierarchy using similarity links or additional static potential functions.

Chapter 8
Scene Understanding for Intelligent Vehicles

Increasing road safety and driver comfort is one of the main research topics for intelligent vehicles. Sensing the environment of the vehicle is the first step within the processing procedure and concurrently the most difficult one due to error-prone and noisy sensor data. Commonly used sensor technologies are radar-based, laser-based, and acoustic-based, which are usually referred to as active sensors. These sensors have in common, that they are transmitting different kinds of waves into the environment and analyzing the reflected wave. This raises a few serious problems for active sensor systems: interferences among sensors of different vehicles, slow scanning speed, low spatial resolution, additional costs and space for the emitter. Passive vision-based sensors, like standard cameras, are capturing data in a non-intrusive way. Since the introduction of inexpensive cameras and of powerful image processing hardware this sensor technology has become attractive for many intelligent vehicle applications like lane detection [113, 145], traffic sign recognition [74, 24], obstacle detection [265, 50], or pedestrian detection [77, 43].

A preliminary version of this chapter was published in [222].

8.1 Related Work

We will now summarize related work, which we divide into general scene understanding approaches and different categories of range data acquisition approaches. These categories include mono, structure from motion, inverse perspective mapping and stereo approaches. Mono-camera approaches, where just one single camera image is used, are often dedicated to explicit applications. Thus, the interpretation is often restricted to the detection of specific objects [218, 79, 253]. Song et al. [218] detected vehicles by finding good vehicle hypotheses using a polar accumulation function which is applied to a virtual top-down view image. The hypotheses are classified using Haar-like features which are learned using the Adaboost learning scheme. Haar features were also used in [79] for pedestrian detection. The Haar features find pedestrian hypotheses which are classified based on histograms of local gradients and support vector machines. A saliency attention mechanism was used in

© Springer International Publishing Switzerland 2015   161
J. Spehr, *On Hierarchical Models for Visual Recognition & Learning of Objects, Scenes, & Activities,*
Studies in Systems, Decision and Control 11, DOI: 10.1007/978-3-319-11325-8_8

[253] to detect road traffic signs. The traffic sign saliency map is constructed using edge and color feature maps. Traffic sign regions are then selected by local maximum energy searching. An evaluation of road marking feature extraction can be found in [236]. The authors systematically evaluated different extraction approaches (positive-negative gradients [111], steerable filters [145], top-hat filters [236], global thresholds, local thresholds, symmetrical local thresholds [72] and color images) using a database of 116 road scene images and found, that the best performances in the general case are obtained by the symmetrical local threshold extractor. Alvarez et al. [5] detected roads in single camera images. They combined low–level cues (appearance of roads) with contextual cues (horizon lines, vanishing points, 3d scene layout and 3d road stages). As a result they got segmented camera images where the road pixels are labeled as one region. Often scene understanding is treated as a segmentation problem. Bileschi [15] segmented street scenes in common classes such as cars, pedestrians, roads and trees using a biologically inspired image representation. A conditional random field was used by Wojek and Schiele [252] to combine object detection and scene labeling within one framework. Traffic scene segmentation was solved by Ess et al. [51] in two stages. In the first stage, they perform an oversegmentation of the image to obtain "superpixels". Then, feature sets are extracted in each segment and a classifier assigns semantic scene labels like road types, cars or pedestrian crossings. Sturgess et al. [225] proposed a segmentation which combines appearance and structure from motion features.

Another common 3d computer vision approach is also based on single camera images but tries to estimate 3d information by means of the underlying scene structure. Delage et al. [40] assumed a "floor-wall" geometry on the scene. 3d reconstruction is solved by a dynamic Bayesian network model, which recognizes the floor-wall boundary in each column of the image. Hoiem et al. [90] tried to recover the surface layout from an image by segmenting it into geometric classes, which coarsely describe the 3d scene orientation of the corresponding region. They learned appearance-based models of these geometric classes based on color, texture, and perspective cues. In [91], they extended their work by integrating occlusion boundaries, objects, camera viewpoint, and relative depth information. Saxena et al. [203] also tried to learn 3d scene structure from single images. Superpixels are used to get a rough segmentation of the scene into small homogeneous patches. For each patch a MRF is used to infer the 3d location and 3d orientation of the patch. Gupta et al. [85] represented the 3d scene by volumes and masses in order to reason about physical constraints within the scene. They used a 3d parse graph which infers object properties like physical boundaries or mechanical properties. Other approaches like [260, 123] grouped lines into surfaces and used them to describe the structure of the scene, or approximated the global depth by identifying of typical 3d scene geometries [164].

Inverse perspective mapping (IPM) [139] is another common 3d vision approach, which is widely used for autonomous vehicles. It uses one single camera image and calculates a top-down view image of the scene assuming a planar ground plane. Kim et al. [110] detected and tracked lanes in top-down view images using the RANSAC approach for efficient selection of lane hypotheses and a particle filter for tracking.

Aly [6] generated lane hypotheses using the RANSAC approach as well. Nieto et al. [168] represented the road by a bipartite graph. The low-level nodes correspond to control points which are detected using a combination of edge detection and morphological filtering. The control points are combined to lane markings, the lane markings to lanes and finally the lanes to the road. The propagation from one level to the next consists of a RANSAC-based estimation method which is applied to analytical functions. Unfortunately, this propagation does not handle uncertainties and the hierarchy is mainly introduced for flexibility enhancement of the system.

Structure from motion approaches (SFM) assume that an image sequence as well as a calibrated camera system (including the relative spatial transformations between the camera positions) is given. It allows 3d reconstruction of the scene by using one camera. In [158] SFM was used to solve SLAM. Their system operates in an incremental way, where in each iteration a new key-frame and 3d points are added. Finally, local bundle adjustment improves the accuracy of the current position. Brostow et al. [26] used SFM to get a semantic segmentation of the images. 3d motion and structure cues (height above the camera, closest distance to camera path, surface orientation, track density and backprojection residual) are projected back to the 2d image plane and a randomized decision forest performs the segmentation.

Similar to IPM, stereo approaches assume that more than one image is given. Different than before, the images are taken simultaneously with a binocular camera setup [87, 206]. Bertozzi et al. [12] used a stereo setup combined with an IPM approach for obstacle and lane detection. IPM is used to get remapped views from the left and right camera and a simple difference operation between these remapped views allows the detection of obstacles. In [23], obstacle detection was solved using stereo images as well. They computed the disparity image, estimated the cameras pitch oscillation, computed the disparity space image and localized obstacles. In [208] road markings were extracted in urban environments using a median local threshold extractor and the authors showed, that optional stereo vision increases the performance of the extraction algorithm.

Although the explicit detection of particular object classes, the 3d reconstruction of the scene or the segmentation are sufficient approaches for many applications, they are not providing a high-level scene understanding regarding spatial, temporal or contextual relations between objects classes. Often these approaches are related to simple applications like lane or pedestrian detection where precise 3d data gathered by laser scanners or stereo setups is available. As soon as the sensor information becomes less accurate or the applications/scenes become more complex, high-level scene understanding has to be used including the integration of uncertainties, high-level reasoning and the combination of several inaccurate sensors readings or preprocessing approaches.

High-level scene understanding is often solved using generative graphical models. Fei-Fei and Perona [129] used a modified version of the Latent Dirichlet Allocation (LDA) model to describe and learn natural scene categories. Unfortunately, the LDA model is a bag-of-words representation which ignores the spatial distributions of the objects. Sudderth et al. [228] explicitly modeled spatial information by adding reference positions for each object. The reference position allows to model the spatial locations of detected features, and thus e.g., to segment street scenes containing buildings, cars, and the road. Other work for the understanding of dynamical scenes include recognition systems from static camera positions where street scenes and intersections are observed from a bird's eye view. Wang et al. [243] used a hierarchical Bayesian model for activities and interactions in crowded and complicated scenes. They exploited the LDA mixture model, the Hierarchical Dirichlet Processes (HDP) mixture model, and the Dual Hierarchical Dirichlet Processes (Dual-HDP) model to connect low-level visual features, atomic activities, and interactions. A similar model was proposed by Hospedales et al. [92]. Kuettel et al. [116] extended the HDP model to find temporal rules by learning sequences of activities. Wojek et al. [251] proposed a generative 3d scene model which combines multi-class object detection, object tracking, scene labeling, and 3d geometric relations. They employed reversible-jump Markov Chain Monte Carlo (MCMC) sampling to estimate 3d scene context and 3d multi-objects. Reversible jump MCMC sampling was also used by Geiger et al. [78] to infer geometric (e.g., street orientation), topological (e.g., number of intersecting streets) and semantic (e.g., traffic situations at an intersection) properties of the scene.

Related to our parking spot finding application is the work [239], where SFM is used. The 3d point cloud generated by SFM is directly interpreted using a model free interpretation as well as a knowledge based interpretation strategy. Car park markings were detected in [107, 106] using a Hough transformation applied to a virtual top-down view. The guideline and the park markings were distinguished by a simple distance metric.

As previously mentioned, approaches which are solely relying on one specific computer vision approach or just using one kind of appearance cue, are expected to perform poorly in real world applications. Therefore, one requirement of the representation must be the ability to combine several visual cues and different low-level vision approaches. This trend can also be seen in the more recently published works [91, 225, 251, 78]. Another requirement of the representation is the ability to integrate uncertainties as well as (spatial, temporal, semantic, contextual) relations between objects (road, cars, pedestrians, etc.). Although approaches like e.g. [228, 91] already model these relations, they are restricted to high abstraction levels and thus cannot benefit from relations at lower levels.

The main contribution of the work proposed in this chapter is threefold. First, we introduce the fusion of different 3d reconstruction approaches to get a robust monocular vision system. The fusion will be performed at different abstraction levels allowing us to combine different approaches, which are sensitive for varying object classes. Furthermore, we propose a scene interpretation method based on a hierarchical representation of scenes and their objects. Scenes are comprised of objects, the objects are in general composed of parts, and these parts are again composed of sub-parts, and so on. We combine the hierarchical decomposition with an efficient sharing structure allowing fast inference and robust detection. Last but not least, in order to make the computational complex belief propagation applicable for intelligent vehicles, we introduce a particle representation in combination with a cluster-based reduction method, make the high-level nodes in the hierarchy observable and demonstrate the efficiency of the proposed representation by means of a parking spot finding application.

The chapter is organized as follows. First, we explain in detail our virtual sensor concept, and describe the structure from motion and inverse perspective mapping approach (8.2.1). Then, the hierarchical scene interpretation is introduced. The virtual sensor data is transformed into occupancy grid maps (OGM) (8.2.2) and interpreted in a hierarchical framework (8.2.3). This interpretation avoids the difficult fusion of low-level information, i.e. the information stored in the OGMs, by introducing several high-level interpretation stages and performing the fusion within these levels. Furthermore, the hierarchical graphical models allow an efficient calculation of the object poses and it is also possible to regard uncertainty caused by sensor noise. We evaluate the performance of the proposed scene understanding in Sec 8.3.1 and demonstrate the interpretation in a parking spot finding application (8.3.2).

8.2 Hierarchical Scene Interpretation

One major aim of intelligent vehicles is autonomous driving in the real world. This requires on the one hand sensors that extract 3d information from the environment and on the other hand a framework allowing the efficient interpretation of the 3d data. Unfortunately, there is no perfect vision-based approach to extract 3d information. Every approach has its pros and cons. For this reason, we introduce the concept of virtual vision sensors. The idea is that different approaches can be applied to the same images of a physical camera. The reconstruction results can be combined to benefit from the pros of every approach.

8.2.1 Virtual Sensors

Each virtual sensor V gathers a type of 3d information using a certain 3d vision approach. Considering one time step t, we get a measurement z_t^V which consists of a point cloud $\{p_i^{z_t^V}\}_{i=0}^{N_{z_t^V}}$. Furthermore, every 3d point is associated with a weight which reflects the uncertainty of the 3d point. In the following we will focus on monocular passive vision approaches. These approaches are the most challenging ones since the only information from the vehicle's surrounding is an image sequence $I_1, I_2, ..., I_t$.

8.2.1.1 Structure from Motion

Using multiple view geometry allows the 3d reconstruction of the scene. Besides the use of multiple cameras it is also possible to use one moving camera and capture images at different time steps, which is called structure from motion (SFM). One very important assumption made is that during the movement of the camera the scene is static. The basic idea is to find point correspondences in the images and

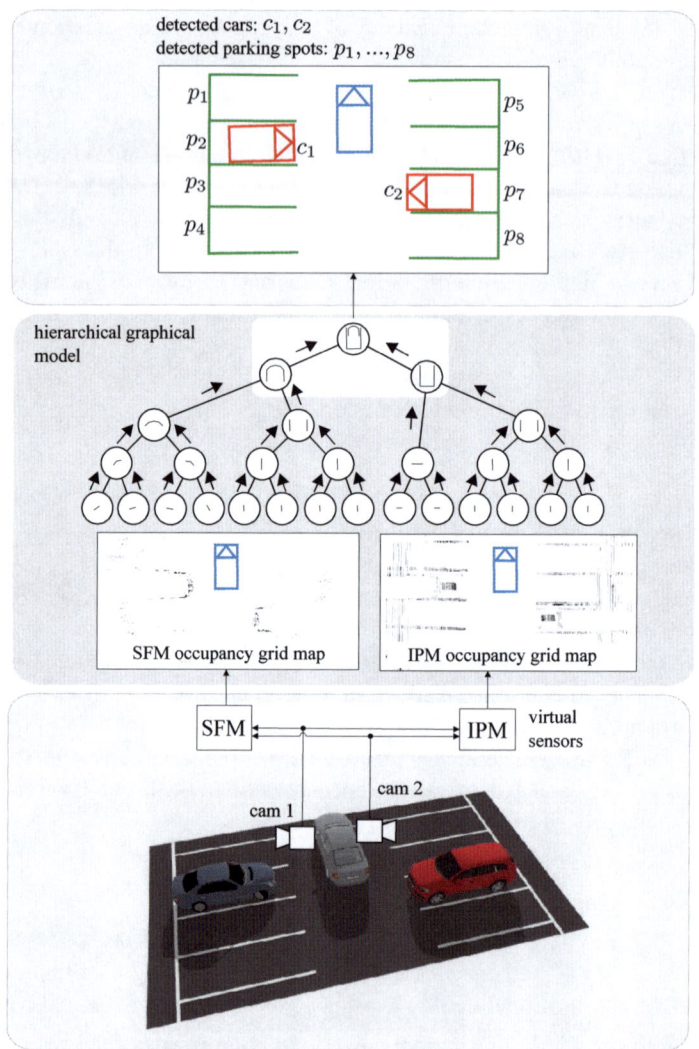

Fig. 8.1 Flow chart of our scene understanding: Cameras capture images of the scene (bottom); the image data is transformed into OGMs using the 3d reconstruction approaches SFM and IPM (center); a hierarchical graphical model combines the information of the two OGMs; the high-level nodes of the hierarchy allow to detect and localize objects like cars and parking spots (top).

extract 3d information by intersecting the related viewing rays in 3d. Feature-based SFM uses edges (like Sobel), corners (Harris corners [86], FAST corners [197, 198]) or other (e.g. SIFT [135]) to define and characterize points in the image and to find point correspondences between them. The search can be done by a feature tracker (e.g. KLT - tracker [137]) or by a direct search along the epipolar line. After point correspondences have been found, the so-called mid-point method can be applied [87]. Thus, 3d points P with closest distance in average to the rays are determined. The amount of all triangulated 3d points at time step t form the measurement z_t^{SFM}.

8.2.1.2   Inverse Perspective Mapping

Using additional information like constraints allows reconstructing 3d information from monocular image data. One very common constraint is that a 3d point is element of a known plane. Then the 3d point of an image point can easily be reconstructed by the intersection of the associated ray with the plane. This so-called inverse perspective mapping (IPM) is a well-known method in computer vision [139]. Applied to intelligent vehicles, we assume that the ground around the vehicle can be approximated by a plane. Thus, we are able to formulate an obstacle detector as follows: First, find possible contact points of obstacles on the ground by means of an appropriate edge detector. Then, reconstruct the 3d points under the assumption that they are elements of the ground plane. Similar to SFM we get a measurement z_t^{IPM} at time t including a set of 3d points and associated weights w_i. The weights are adjusted w.r.t. to the response of the edge detector result and are normalized.

8.2.2   Low-Level Scene Representation

8.2.2.1   Occupancy Grid Maps

The vehicle's environment is represented using OGMs which are widely used in mobile robotics. The idea of this approach is that the environment is represented by metric grid cells. Every grid cell (x, y) has an occupancy value which usually reflects the probability that the grid cell is occupied by a static obstacle. The virtual sensors are used to estimate the probability values of the OGMs. A general virtual sensor delivers the measurements $z_1, ..., z_T$ from time 1 to time T. Using these measurements we want to obtain the OGM o according to $p(o|z_1, ..., z_T)$. This idea was first introduced by Moravec and Elfes [153] and enhanced by Thrun et al. [232] and others. We assume that the environment is static and each grid cell of the map $o_{x,y}$ is conditional independent of its neighboring cells. To reduce the computational cost log-odds of $p(o_{x,y}|z_1, ..., z_T)$ are used, which can be calculated using

$$\log \frac{p(o_{x,y}|z_1, ..., z_T)}{1 - p(o_{x,y}|z_1, ..., z_T)} = (T - 1) \log \frac{1 - p(o_{x,y})}{p(o_{x,y})}$$
$$+ \sum_{t=1}^{T} \log \frac{p(o_{x,y}|z_t)}{1 - p(o_{x,y}|z_t)} \quad (8.1)$$

$$l_{x,y}^T := \log \frac{p(o_{x,y}|z_1, ..., z_T)}{1 - p(o_{x,y}|z_1, ..., z_T)} \tag{8.2}$$

With the log-odds $l_{x,y}^T$ the posterior occupancy probability becomes

$$p(o_{x,y}|z_1, ..., z_T) = 1 - \frac{1}{1 + e^{l_{x,y}^T}} \tag{8.3}$$

After solving the induction over t it can be shown (see [232] for a detailed overview)

$$l_{x,y}^T = (T - 1) \log \frac{1 - p(o_{x,y})}{p(o_{x,y})} + \sum_{t=1}^{T} \log \frac{p(o_{x,y}|z_t)}{1 - p(o_{x,y}|z_t)} \tag{8.4}$$

The probability $p(o_{x,y})$ contains prior information and the probability $p(o_{x,y}|z_t)$ reflects the occupancy of the grid cell $o_{x,y}$ conditioned on the measurement z_t. This conditional probability is determined by the measurements of the virtual sensors. First, the 3d points are transformed into the occupancy map coordinate system using a homogeneous transformation F_t. Initially, F_0 can be arbitrarily chosen. In every time step F_t has to be updated using odometry data provided e.g. by the ego motion of the vehicle or some visual odometry approach.[1] Then, every transformed point increases the probability that the associated grid cell is occupied. We use a voting table vt with the same size and resolution as the OGM, and count the transformed points in every grid cell: $vt_{x,y}^{z_t} = \sum_{i=0}^{N^{z_t}} w(i)\delta(F p_i^{z_t} - (x, y, 1)^T)$, where $\delta(x) = 1$ for $x = 0$, and $\delta(x) = 0$ else. The probability $p(o_{x,y}|z_t)$ is calculated using

$$p(o_{x,y}|z_t) = 1 - e^{-(vt_{x,y}^{z_t})^2/(2\sigma^2)} \tag{8.5}$$

where the parameter σ regards the presence of noise. At this point we extend the idea of OGMs to any kind of object class, not just static physical obstacles. Such an object class could be the class of dynamic obstacles, road or park markings. Every object class C gets its own OGM $p(o^C|z_1, ..., z_T)$ which contains the probability that a certain cell is occupied by an object of the associated class. The problem with that is, that there is no perfect sensor, which is just sensitive for a particular object class. IPM uses e.g. an edge detector and therefore is sensitive for obstacles and non obstacles, like markings. Thus, we would have to estimate $p\left(o^C|z_1^{V_1}, ..., z_T^{V_1}, ..., z_1^{V_n}, ..., z_T^{V_n}\right)$ of the OGM based on the measurements gathered by all virtual sensors. Unfortunately, due to the complex dependencies, which also exist between neighboring cells, we cannot simplify this problem to $p\left(o_{x,y}^C|z_1^{V_1}, ..., z_T^{V_1}, ..., z_1^{V_n}, ..., z_T^{V_n}\right)$, which would allow a low-level combination of the measurements. We avoid the problem by introducing OGMs for every virtual

---

[1] Practically, we choose vehicle centered OGMs where in each time step the relative pose of the OGMs is updated.

sensor. Thus, the OGMs can be calculated as $p\left(o_{x,y}^V|z_1^V,...,z_T^V\right)$ using the log-odds representation. The fusion of the OGMs is integrated in a hierarchical interpretation framework. As we will see, this allows to handle the complex dependencies between the maps occuring due to occlusions, sensor properties, obstacle features and so on.

Fig. 8.1 (center) shows an example of two OGMs gathered by a SFM and a IPM virtual sensor. The obstacle points gathered by the SFM virtual sensor show the two U-shape like obstacles representing two parking cars. The non obstacle points detected by the IPM virtual sensor show primarily the park markings.

8.2.2.2 Preprocessing

The occupancy grid cells just provide local occupancy probabilities. Furthermore, we intend to use local information, like e.g. the local gradient or the orientation. For that, we apply an additional preprocessing step to the OGMs. A window of size (w_x, w_y) is moved over the maps, where the step width during the movement is $(w_x/2, w_y/2)$. At every window position a local normal distribution is calculated for the occupancy grid cells within the window. The mean of the resulting distribution is the center point $(x, y) \in \mathbb{R}^2$ and the largest eigenvector of covariance matrix allows to approximate the local orientation α and the width l. Furthermore, we can use the relation of the eigenvalues λ_1/λ_2 to derive a score how unambiguous the result is. This information defines a line element $\boldsymbol{y}_i = (x_i, y_i, \alpha_i, l_i) \in \mathbb{R}^4$, which will be used as low-level information of the hierarchical interpretation.

8.2.3 Scene Understanding

The detection of objects like cars, markings or obstacles is very challenging due to the large number of object classes, their unknown number in the scene, their unknown pose and their unknown object parameters. The proposed scene interpretation aims to infer the high-level information based on the low-level sensor input in an efficient manner using a hierarchical object representation. So instead of searching the whole parameter space for an arbitrary object, we first look for its sub-parts and make the detection result available to all parent parts. This propagation scheme is very efficient since redundant calculations are avoided.

Furthermore, the hierarchy allows to combine the OGMs of the different virtual sensors. The SFM OGM will mainly contain obstacles like cars, the IPM OGM will on the other hand mainly contain markings. Thus, the fusion of the OGM at low-level is very difficult. However, the spatial dependencies between parked cars and parking spots are very obvious, the parked car will generally stand between the markings. We will use these spatial dependencies in order to combine the OGMs of the different virtual sensors.

The OGMs are the lowest level of our hierarchical framework. On the next higher level we are using line objects. Line objects are part of many objects e.g. walls, cars or markings. These line objects can now be combined to form arc shapes and parallel lines. One arc shape and two parallel lines form an U-shape, and so on.

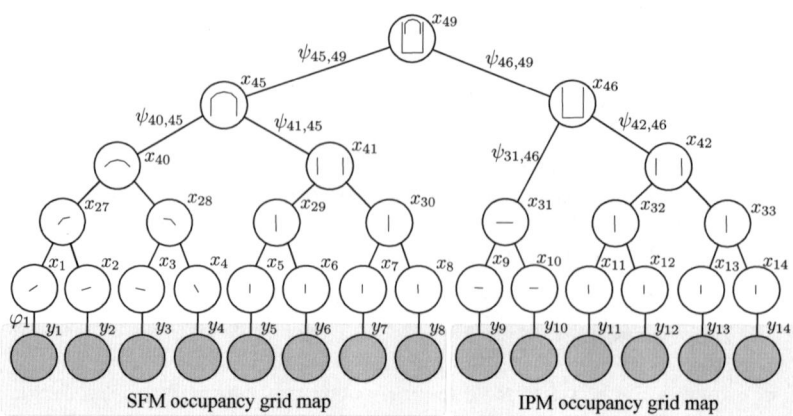

(a) Hierarchical structure of an occupied parking spot.

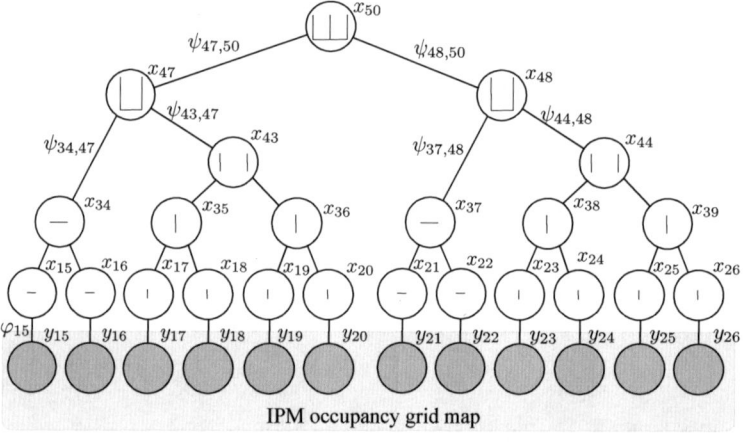

(b) Hierarchical structure of a parking place with two spots.

Fig. 8.2 Hierarchical structure of the part and sub-part decomposition. The nodes of each level have a similar complexity and are combined to form the parent part of the next higher level. The correlation functions $\psi_{ij}(\boldsymbol{x}_i, \boldsymbol{x}_j)$ define the spatial relation between two hidden nodes. The symbols within the hidden nodes represent object primitives like line elements (e.g. node $\boldsymbol{x}_1$), arc shapes (e.g. node $\boldsymbol{x}_{27}$) or parallel lines (node $\boldsymbol{x}_{41}$) and objects like cars (represented by an U-shape, node $\boldsymbol{x}_{45}$) or a parking place with two spots (node $\boldsymbol{x}_{50}$).

The model is represented by set of hierarchies $\mathcal{G} = \{\mathcal{G}_1, ..., \mathcal{G}_{N_\mathcal{G}}\}$, which are specified by undirected tree-structured graphs $\mathcal{G} = (\mathcal{V}, \mathcal{E})$ as described in Sec. 3.7. The graph $\mathcal{G}$ is composed of hidden random variable nodes $\boldsymbol{x} = \{\boldsymbol{x}_1, ..., \boldsymbol{x}_M\}$ and observed nodes $\boldsymbol{y} = \{\boldsymbol{y}_1, ..., \boldsymbol{y}_N\}$. The hidden random variable nodes represent the object, a part of an object or a sub-part of an object. In general, the nodes have the form $\boldsymbol{x}_i = (x_i, y_i, \alpha_i) \in \mathbb{R}^3$, where (x_i, y_i) are the OGM coordinates and α_i is an orientation angle. Of course other dimensions concerning object parameters like

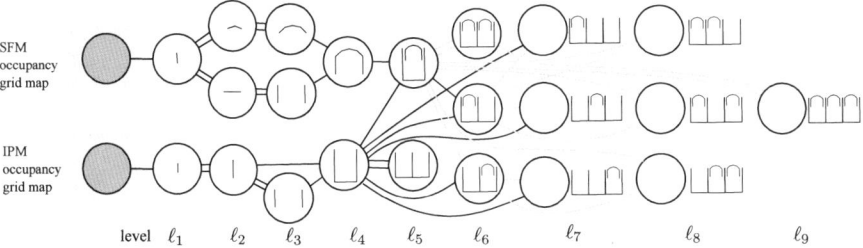

SFM
occupancy
grid map

IPM
occupancy
grid map

level ℓ_1 ℓ_2 ℓ_3 ℓ_4 ℓ_5 ℓ_6 ℓ_7 ℓ_8 ℓ_9

Fig. 8.3 Sharing structure used during message passing to avoid redundant calculations and to increase the performance.

curvature, width or height can be added. The observed nodes represent the line elements $\boldsymbol{y}_i = (x_i, y_i, \alpha_i, l_i) \in \mathbb{R}^4$ that are obtained in the preprocessing step (see Sec. 8.2.2.2). The nodes represent the leaves of our hierarchical representation. An example of the representation can be seen in Fig. 8.2. Since we are interested in a high-level representation of the scene our aim is to propagate the low-level OGM information to high-level object information. This can be efficiently done using belief propagation [179, 259], where message passing is applied to propagate the information from the leaves to the root node. Nevertheless, due to our continuous-valued graphical model belief propagation is computationally very complex and hence an inappropriate means for a scene interpretation framework for intelligent vehicles. We therefore use the nonparametric belief propagation framework as described in Sec. 2.3.2.

Sharing Parts: The sharing of parts can significantly increase the performance of the message passing as already discussed in Sec. 3.7.3. It is furthermore one of the main reasons for the hierarchical decomposition into generic geometric elements like lines, arcs or U-shapes. During the upward-sweep, the belief of the short line elements are calculated just once. The results are sent to all parents nodes, which are the arc elements and longer line elements. Similar, at higher levels objects like single parking spots are calculated just once, and then combined to larger places with two parking spots, three ones and so on. This sharing avoids redundant calculations and thus accelerates the belief propagation (see Fig. 8.3).

Reduced Particle Sets: We reduce the sample set using a clustering based sample reduction method as described in Sec. 3.8.1.1. The idea of this approach is to cluster the samples and employ the cluster centers as the reduced dataset. Joen [102] proposed to use the Isodata clustering procedure. This method is similar to the k-means clustering, but it overcomes the drawback of a fixed number of clusters by removing redundant clusters. However, an initial maximum number of clusters still has to be chosen. We use the density-based clustering method [52] which requires just two parameters: the maximum inner-cluster distance and the minimum number of samples per cluster (see Sec. 3.8.1.1).

Observable High-level Nodes: Until now, we assume that sensor information is fed into the model via low-level nodes. However, introducing additional

observable nodes y for the high-level nodes x can significantly improve the accuracy of the localization. The idea is further described in Sec. 3.5.2. This step is
similar to the particle filtering approach, where a deterministic drift and a stochastic
diffusion is applied to the particles and finally the particles are weighted using the
observation function. The only difference between the two approaches is, that we
use the weighting step at each level instead of each time step as the particle filter
does.

Periodic Variables: At least one dimension of the random variables represents a
periodic variable, which is conveniently represented using an angular coordinate
$0 \le \alpha < 2\pi$ [16, 226, 140]. Commonly, this dimension corresponds to the rotational part of the variable nodes x. In order to avoid problems concerning the
choice of the origin, we model the angular dimensions by a von Mises distribution
$\mathcal{M}(\alpha; \mu, \kappa) = (2\pi I_0(\kappa))^{-1} \exp\{\kappa \cos(\alpha - \mu)\}$, where the concentration param-

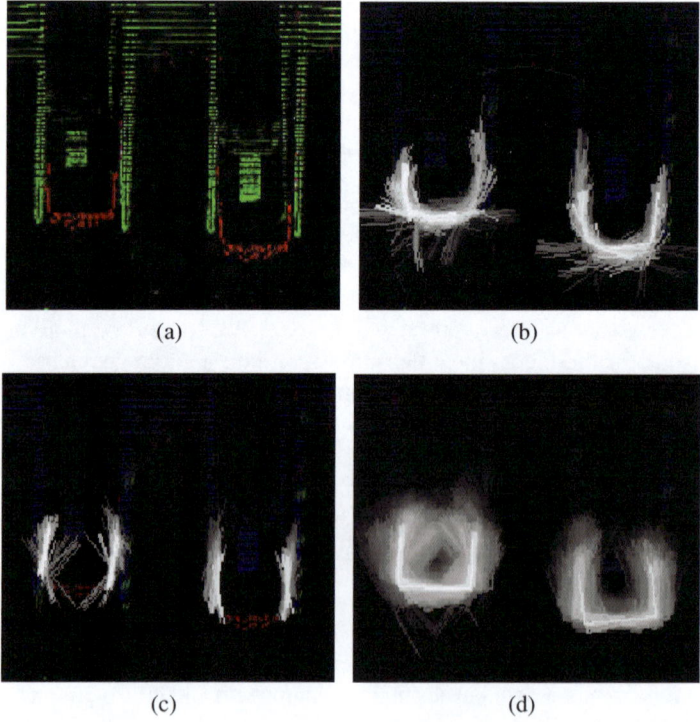

(a)   (b)

(c)   (d)

Fig. 8.4 Example for the beliefs $b_i(x_i) = p(x_i|y)$ of different levels in the hierarchy. The
images show a scene containing two cars (a); the SFM OGM (red) and the IPM OGM (green).
The PDFs are represented by particles which are shown in Figure b-d. During the buttom-
up propagation the low-level information of the obstacle occupancy grid cells and the line
elements (a) are combined to arc shapes (b), the line elements to parallel lines (c), and finally
the curved front and the parallel lines to U-shapes (d).

eter κ corresponds to the inverse variance σ^2 of a Gaussian: $\kappa \approx \sigma^{-2}$. $I_0(\kappa)$ denotes the zeroth-order Bessel function of the first kind. The von Mises distribution is particularly suitable for our framework since it can be derived from a bivariate Euclidean Gaussian distribution [226]. We can thus represent the von Mises distribution by a Gaussian in the 2d Euclidean space and constrain the samples to lay on the unit circle.

8.3 Results

8.3.1 Performance and Accuracy

The main questions for the application of the hierarchical framework in an intelligent vehicle are, how accurate can the high-level objects be recognized and whether the framework is real-time applicable. We evaluated the scene interpretation in a calibrated test environment, where OGMs and ground truth data were available. Here, we used our framework to estimate the position of a car in the scene represented by an U-shape, as can be seen in the left branch of Fig. 8.2(a). The line parts are defined by $x_1 = (x_1, y_1, \alpha_1, l_1) \in \mathbb{R}^4$, where (x_1, y_1) are the OGM coordinates, α_1 is an orientation angle and l_1 is the length of the line. On the next higher level, short arc shapes are used to model local curvature. They have an additional curvature angle ϕ, $x_{27} = (x_{27}, y_{27}, \alpha_{27}, l_{27}, \phi_{27}) \in \mathbb{R}^5$. The short arc shapes are combined to longer and more complex arc shapes which represent the front of the car, $x_{40} = (x_{40}, y_{40}, \alpha_{40}, l_{40}, \phi_{40}) \in \mathbb{R}^5$. Furthermore, the sides of the car are represented by two parallel lines; they are defined by $x_{41} = (x_{41}, y_{41}, \alpha_{41}, l_{41}, w_{41}) \in \mathbb{R}^5$, where w_{41} is the distance between the lines or the width of the car. Finally, the curved front and the parallel lines are combined to an U-shape of a car defined by $x_{45} = (x_{45}, y_{45}, \alpha_{45}, l_{45}, w_{45}, \phi_{45}) \in \mathbb{R}^6$. The potentials $\psi_{ij}(x_i, x_j)$ are modeled by a diagonal Gaussian conditional distribution, where the potentials between the line elements x_1 and the arc elements x_{27} is exemplarily

$$
\begin{aligned}
\psi_{1,27}(x_1, x_{27}) = &\mathcal{N}(x_{27}; x_1 + s_1 \cos(\alpha_1), \sigma_p^2) \times \\
&\mathcal{N}(y_{27}; y_1 + s_1 \sin(\alpha_1), \sigma_p^2) \times \\
&\mathcal{N}((\alpha_{27}, \phi_{27}); (\alpha_1, 0), \Lambda_\alpha) \times \\
&\mathcal{N}(l_{27}; l_1, \sigma_l^2)
\end{aligned}
\tag{8.6}
$$

The other potential functions have a similar structure, but cannot be explained in detail. The potentials are augmented by a zero mean, high-variance Gaussian, weighted to represent 25% of the total likelihood. This allows handling outliers due to occlusion. As can be seen in Fig. 8.5 (left) the recognition performance is almost the same for different setups. However, Tab. 8.1 shows, that especially the reduction of the particles sets accelerates the belief propagation significantly. Furthermore, introducing additional high-level observation nodes improves the accuracy of the localization, as can be seen in Fig. 8.5 (right). This improvement makes the interpretation

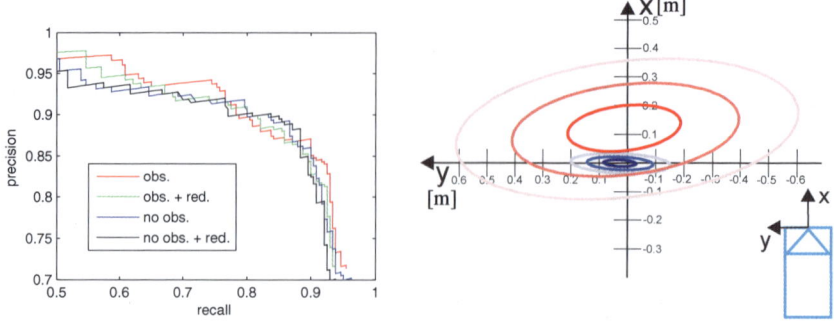

Fig. 8.5 Results: Recall precision curves for different setups - with observable (obs.) high-level nodes and with reduced (red.) particle sets (left); localization error modeled as a normal distribution with (blue) and without (red) additional high-level observation nodes (right).

Table 8.1 Complete execution time for different setups

sharing	yes	yes	yes	yes	no
observation	no	no	yes	yes	no
reduction	no	yes	no	yes	no
time (ms)	23.93	8.27	57.57	9.00	71.80

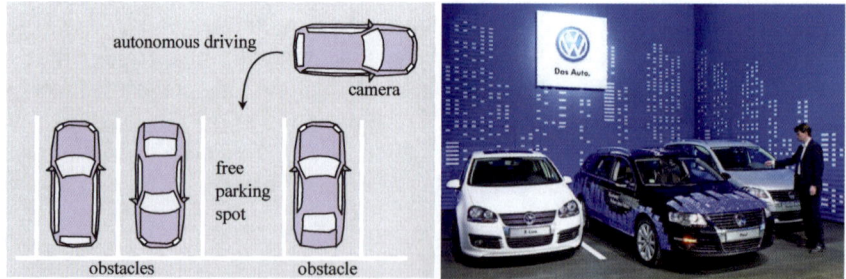

Fig. 8.6 Parking spot finding application (left); Experimental vehicle Paul (right) (image source: Volkswagen AG 2008).

applicable for real-time applications. Fig. 8.4 shows samples of the different levels during the recognition step.

8.3.2 Parking Spot Finding Application

We demonstrate our approach using a parking spot detection application. Considering a vehicle driving along parking spots on the right and on the left, our aim is to detect with a monocular vision system those, which are free. For that, we equipped the vehicle with two cameras, one under the left side mirror looking to the left, and

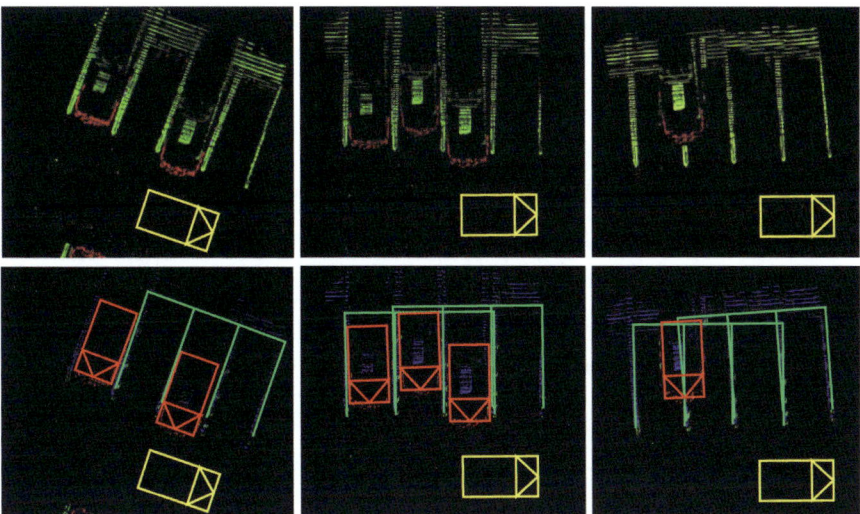

Fig. 8.7 Three results of the parking spot finding application: (top) Input OGMs. (bottom) Ego (yellow), detected parking spots (green), and detected cars (red).

one under the right side mirror looking to the right (Fig. 8.1). Please note, the two cameras are only used to enlarge the registration area. The virtual sensors are still based on monocular approaches.

The cameras have wide angle lenses and can cover a large area of the vehicle's lateral space. Due to the finite exposure time of the cameras the vehicles velocity during the parking spot detection is limited to 20 km/h. The image data is processed according to the virtual sensor concept. The SFM-based virtual sensor uses a real-time Sobel feature reconstruction algorithm. First, images of two time steps are rectified using look-up tables and vertical image features are extracted. Then, point correspondences are found on the epipolar lines. To calculate the matching cost of the point correspondence, a normalized cross correlation is used. The relative transformation between the two time steps is estimated using the odometry data of the vehicle. The IPM-approach uses Sobel features, too. We use OGMs with a grid cell size of 5cm × 5cm.

The interpretation of the OGMs is performed using our hierarchical scene interpretation framework of car objects and the parking place. The potential functions associated to the edges of the hierarchy were manually defined. In order to decrease the computation time of the approach, additional constraints concerning the orientations of the parking place and the car orientation were made. Thus, we are able to run the whole image processing and interpretation with 20fps on a 2.4GHz double core processor, where image capture and image processing are separated in two threads.

Free parking positions are identified based on the recognized parking places and cars. First, we search for spots of the parking place which are not occupied by a car.

Since it is possible that there are parking places without marking, we also search for free parking spots beside recognized cars. We are able to achieve a detection rate of about 90.5% and a false positive rate of 5.8%. False negatives are mainly caused by situations, where the ground plane can not be approximated by a plane. Results are shown in Fig. 8.7. The top row represents the input OGMs (SFM (red) and IPM (green)) and the bottom row the corresponding recognition result. We have demonstrated our experimental vehicle Paul at the Hannover fair 2008 where we have shown the reliability of the proposed approach in hundreds of live demonstrations (see Fig. 8.6).

Chapter 9
Conclusion

We developed a new hierarchical representation which is adaptable for a wide range of applications. Our approach exploits three concepts: compositional hierarchies, coarse-to-fine hierarchies, and the sharing of parts to get a robust and efficient representation of multi-objects, multi-scales and multi-views. The applicability of the proposed methods and representations are confirmed by the results achieved in different applications. Our efficient calculation of products of Gaussian mixtures shows that it outperforms standard approaches like exact, importance or Gibbs sampling and is especially suited for object recognition tasks. Furthermore, we show how our combined bottom-up and top-down propagation scheme gets better recognition performance with less particles compared to standard bottom-up propagation. The coarse-to-fine inference makes our representation very fast and robust. Even with just one additional level of detail the performance is significantly better than without the similarity hierarchy. The recognition performance as well as the computational efficiency increases. The model can better handle cluttered backgrounds, occlusions and sensor noise. We also showed that the framework can easily be adapted to pose estimation tasks such as human pose estimation, and can be used for important applications like gait analysis.

Our framework provides a very generic way to represent hierarchies. This is the reason, why we easily can adapt it to problems like human behavior analysis. We just had to change the low-level features and potential functions from 2d image space to the 3d spatiotemporal volume. The results show that for sequences of actions and activities, the hierarchical representation outperforms representations like bag-of-words models where the spatiotemporal dependencies are ignored. The results furthermore confirm the robustness of the hierarchical representation against temporal scaling, noise and occlusions. Most challenging was a long term dataset of activities of daily living. The unsupervised learning allows to build sequentially efficient representations of complex behavior patterns. The learning was able to share primitives up to level 14, which corresponds to activities like 'Entering the room and food preparation'. Although the models and the learning algorithm already provide promising results, there is still a lot to do. Especially, the large variations of the training instances have proven to be challenging.

© Springer International Publishing Switzerland 2015   177
J. Spehr, *On Hierarchical Models for Visual Recognition & Learning of Objects, Scenes, & Activities,*
Studies in Systems, Decision and Control 11, DOI: 10.1007/978-3-319-11325-8_9

The scene understanding for intelligent vehicles is another example, where the hierarchical model has shown to be a good choice. The ability to combine different 3d computer vision approaches at different abstraction levels leads to a robust detection and recognition framework. Furthermore, the sharing and the reduction method decrease the runtime, and the use of additional observable high-level nodes increases the accuracy of the approach. We were able to integrate the 3d scene reconstruction (SFM and IPM) for two cameras (left and right) and the hierarchical framework for real-time with 20fps on a standard computer. Although the park spot finding application covers just a small part of autonomous driving, we see great potential that the new hierarchical model can become a standard model for the efficient representation and perception of scenes for intelligent vehicles.

The objectives of our learning framework differs substantially from other approaches. Our unsupervised learning builds hierarchical models from few or single instances as they are e.g. provided by pose collections. During training the algorithm tries to maximize the reusability of parts and primitives. We propose an online and offline learning scheme and the results show that the offline version is especially suited for learning of multi classes where the search for similar parts is more complex. If the sets contain similar instances, e.g. due to rotational symmetry, the online learning was able to reuse already learned features more efficiently. With minor adaptation effort the learning framework was also applied to the learning of behavior patterns, where efficient representations of complex motion sequences were built automatically.

9.1 Extensions and Future Research

In the following we will give an overview of extensions, improvements and new directions of future work. We also provide a few new areas, where the proposed hierarchical framework could be used to handle existing challenges. General possible extensions were already summarized in Sec. 3.4, where the information sources, which a high-level node in the hierarchy in principle can assess, were described. Although most of the sources were already integrated in our framework, there are still a few, which might also be reasonable. These are:

- **Temporal Tracking**
  Up to now, the detection algorithms described here are not using temporal tracking. Although the activity recognition framework models the temporal relations between actions, it still does not use temporal tracking. The idea of temporal tracking is, that we regard dynamics in the scene, i.e. dynamic objects or a moving camera at discrete time steps. We can model the temporal stochastic process using a HMM of order one. Thus, each node in the hierarchy has a belief estimate at different time steps t and $t-1$. At time step t we can assess additional information provided by the previous time step $t-1$. This information is encoded in the message node x_t receives from its predecessor x_{t-1}. It corresponds to the prediction

$$p(x_t|y_{1:t-1}) = \int p(x_t|x_{t-1})p(x_{t-1}|y_{1:t-1})dx_{t-1} \qquad (9.1)$$

which uses the dynamic model $p(x_t|x_{t-1})$ underlying object x_t. The standard so-lution to integrate this prediction in a probabilistic framework is to use $p(x_t|y_{1:t-1})$ as a proposal function in a sequential importance sampler. This however corre-sponds to a pure tracking method, so that the approach suffers under common tracking problems such as the need for additional initialization routines and the risk of losing the tracked object. We therefore suggest to combine the prediction with the bottom-up and top-down message passing as described in Sec. 3.8.2. For that one could additionally initiate the message passing not just solely based on the observed nodes but also based on the prediction from the previous time step.

- **Global Context**
  Currently, context is just provided locally by the parent node. One could addition-ally use global features (see Sec. 3.5.1) to get global information about context, in which specific nodes (representing an object, or activity) are likely to occur. These global features could be estimated using e.g. gist features, which are a statistical summary of the spatial layout properties of the scene [172, 173].

- **Sharing Attributes**
  The random variables of our object and human pose estimation just model the spatial distribution of the associated objects or parts. For activity representation a temporal dimension was added. For the scene interpretation we also added object properties like width, height, or curvature. We found that the design of the covariance matrices is very challenging since the mutual influence has to be defined. Another promising solution would be to use an additional random variable which represents the attributes of an object. Since attributes are generally the same for an object and its parts, one could share the attributes between all elements of an object-specific hierarchy. For object recognition, an additional attribute variable could for example represent the color values of an object, and this variable could be connected to all elements in the hierarchy associated to the object.

We now give an overview of extensions related to our applications:

- **Object Recognition**
  One problem of the hierarchical representation is that the number of elements increases exponentially with the number of hierarchy levels. Sharing was used to reduce this complexity. In order to further improve the representation, more in-variant features might be helpful. For instance, the similarity measure could also regard affine transformations, which would increase view invariance of parts and thus their reusability. This view invariance could also be achieved by 3d infor-mation. Up to now, objects were completely modeled in image space according to their 2d view and articulation dependent appearance. An interesting extension would be to use 3d features and their 3d relations similar to the hierarchical model proposed by Detry et al. [41]. Unfortunately, these authors did not integrate shar-ing of object parts in their framework. Sharing of 3d primitives, 3d shapes and 3d parts, however, might be a promising research direction as confirmed by our

activity recognition results, which can be seen as an 3d object recognition framework in (x,y,t)-space.

- **Human Pose Estimation**

  In this monograph we used the hierarchical representation of the human body for gait analysis. Due to this specific application we were just interested in the detection of the head, the body, and the legs. The model is of course also suitable for the representation of the whole human body. Here, we expect the hierarchical structure to be even more efficient, since the different arm configurations can be shared (left, right) and, in addition, visual primitives can also be shared with legs. Furthermore, the integration of tracking information (see above) could be used to make the pose estimation more robust over time [214].

- **Human Behavior Analysis**

  Especially, the large variations of the training instances have proven to be challenging for the efficient learning of human behavior patterns. We think there are two possible improvements. On the one hand the hierarchy currently just models the relative spatiotemporal relations. Thus, absolute information about places like the stove, where the activity 'Cooking' usually will occur, are ignored. We assume that introducing absolute spatial and also temporal information would increase the recognition performance. On the other hand, one could reduce the variations by using more invariant features. The current representation models the flow field observed in the camera image. Unfortunately, actions like 'Going from the table to the stove' and 'Going from the door to the stove' will have very different flow fields, although their actual semantical intention 'Going to the stove' is the same. Thus, as an improvement one could choose the reference position of an action as the target position and also use a new distance measure depending on the target position.

- **Scene Understanding for Intelligent Vehicles**

  We applied the hierarchical framework to a parking spot finding application and demonstrated the efficiency and robustness of the approach, which is able to combine different sensor sources at different abstraction levels. The approach is also appropriate for a general understanding of traffic scenes. The different kinds of roads (one, two or three lanes) including different kinds of markings can efficiently be represented by a hierarchy and a corresponding sharing structure. In Fig. 9.1 (left) some simple models of two-lane roads are depicted. As in the parking scene, sharing can be used to increase robustness and performance (Fig. 9.1 (right)). The hierarchical representation is also attractive since a wide range of sensors and detectors already exists. These sensors provide low-level information like distance values as well as high-level information like trajectories of pedestrians. The hierarchical model is an appropriate means to integrate these different kinds of information into one unified probabilistic framework.

The proposed representation of objects, scenes, and activities as sets of hierarchies, which are connected in order to increase robustness as well as efficiency, has many other areas, where an application and adaption might be reasonable. Actually, every problem domain, which can be hierarchically decomposed into simple primitives, is appropriate. In the following we will give examples for promising future directions:

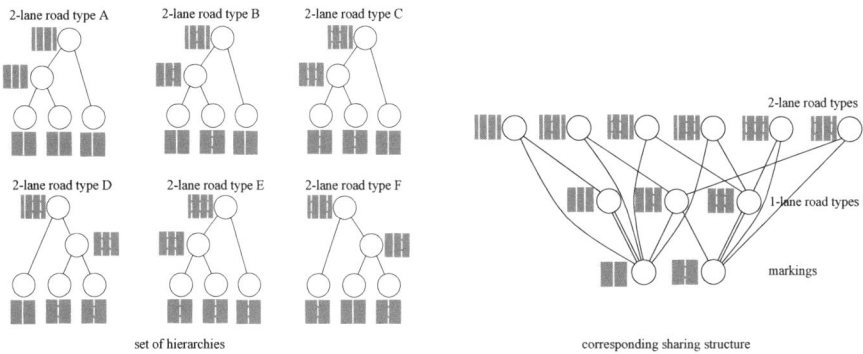

Fig. 9.1 Hierarchical models for different road types: 2-lane types (left) and the corresponding sharing structure (right).

- **Stereo Matching**
  Stereo matching is the well-known problem of finding correspondences between two images in order to reconstruct depth information. There are two basic concepts to solve the correspondence problem. Local methods are trying to find the best correspondence by maximizing the similarity for each pixel separately. Here, area-based similarity measures like normalized cross correlation are used. Global methods are additionally using global constraints such as smoothness, and are generally formulated as an energy minimization problem (e.g. graph cuts, or belief propagation). While global methods are reaching higher quality, they are computationally much more expensive. The idea of using the proposed hierarchical representation for stereo matching is depicted in Fig. 9.2. First, the left camera image is used to learn a hierarchical representation by decomposing the image top-down. Then, the inference algorithm is used to detect the learned model (of the left camera image) in the right camera image. The detected objects, parts and primitives and their position in the right camera image give disparity information. The hierarchical model is attractive since it combines local evidence with global context at different abstraction levels. Thus, the features at lower levels can be aligned according to the image information while the context of the parent node provides the spatial arrangement. The smoothness of the objects is thus guaranteed. Furthermore, the approach does not use a search along the epipolar line, so that it is robust against changed epipolar geometries due to uncalibrated camera systems.
- **Simultaneous Localization and Mapping**
  Simultaneous localization and mapping (SLAM) is used by mobile robots or autonomous vehicles to build up a map of the environment while simultaneously localizing itself within this map [232]. This is a challenging problem since the map as well as the poses of the robot are unknown and additional problems like loop closing have to be solved. We propose to use the hierarchical model described in this monograph for solving the SLAM problem. The idea is to decompose a map into its parts (like e.g. part of town, streets or corridors), and these parts again

left camera image   right camera image

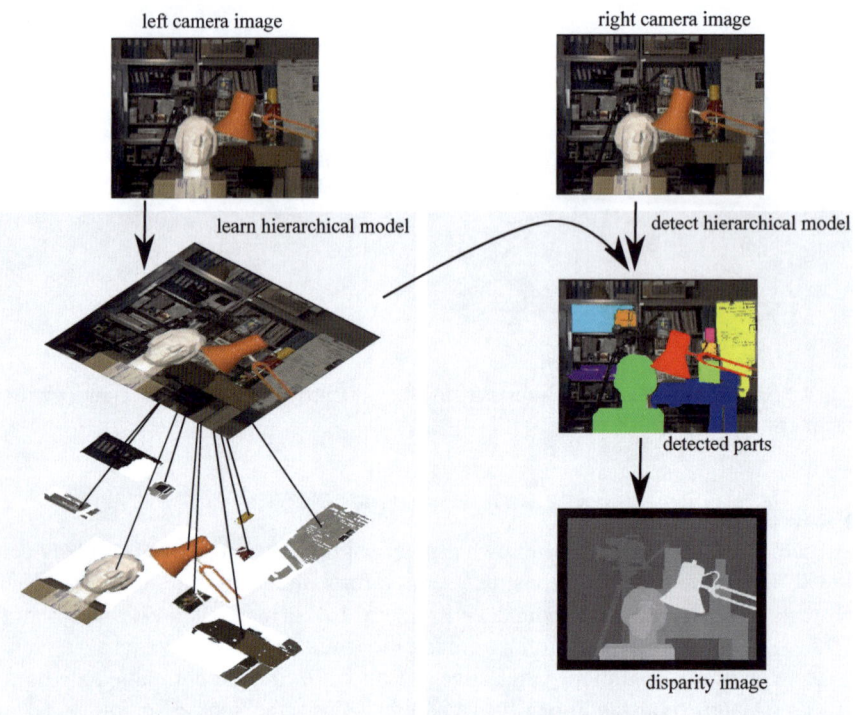

Fig. 9.2 Hierarchical models for stereo matching: the left camera image is used to learn a hierarchical representation by decomposing the image top-down (left), the inference algorithm is used to detect the learned model in the right camera image. The detected objects, parts, and primitives and their position in the right camera image give disparity information (right).

in elements like walls or doors, the elements in smaller elements and so on. An example can be seen in Fig. 9.3(a), where at the highest level a map of an indoor scenario is shown. This representation has several advantages. Let us assume that the elements at the lowest level correspond to local maps which are for example captured by a laser scanner. During map building the relative transformations between the local maps can just be estimated up to some degree of uncertainty. The problem of loop closing is, that the uncertainty increases as more and more local maps are merged. The hierarchy intrinsically solves this problem since it models the uncertainties by means of potential functions, which represent the spatial relations relative to the parent elements. Thus we can increase the variance of the potential functions with the level to guarantee that the local maps are firmly connected while the maps at higher levels are loosely connected. Furthermore, the sharing of different parts reduces the complexity of the map and also improves the localization accuracy. In many scenes, the local maps are actually very similar. This is especially the case for indoor scenes with long corridors. The sharing structure (see Fig. 9.3(b)) models this similarity by reusing the local maps at higher levels. The local maps can for example be combined to a corner and this corner can be reused four times in the map (Fig. 9.3(b)red). The standard way to localize a robot within the map is by means of a matching algorithm. If we assume the robot has a local scan, which corresponds to the previously mentioned corner, we would ideally assume that the matcher would deliver the four corners in the map. However, our hierarchical representation intrinsically encodes this uncertainty by means of sharing the 'sub map' corner, so that the matching algorithm becomes unnecessary.

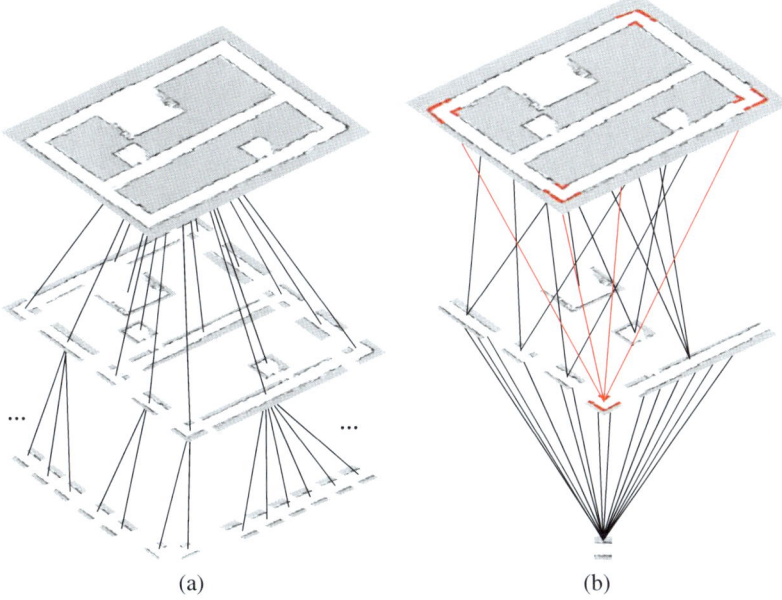

(a)   (b)

Fig. 9.3 Hierarchical models for the SLAM problem: (a) Hierarchical decomposition of a map into map parts, and local maps. (b) Sharing structure (details see text).

- **Human-Machine-Interaction**
  The hierarchical activity model provides some interesting applications for human-machine-interaction. The activity recognition framework allows the spatial and temporal recognition of action primitives, more complex actions, and activities. It thus supports information about the current actions, its context and can also be used to predict the next most likely actions and movements. This induces some important properties, which are needed for intuitive human-machine-interactions. The detected actions and their context can be used by a machine to react according to the context and the most likely predicted action. Challenging is the mapping of the different actions observed on the sensor side to actions on the actuator side. Reasonable seems here the use of learning by demonstration. In a smart home one could for example learn motion patterns of the user and

integrate the use of electronic devices by demonstrating the use. When a pattern such as 'Go into the living room', 'Grasp a book', 'Sit down', 'Switch the reading lamp on', and 'Start reading' were observed several times, the system could learn the pattern and automatically switch the light on, if the context is detected and the current observed action is 'Sitting down'. In a more complex interaction a robot could learn the assembly of an object and react depending on the observed actions. During the assembly the user has to demonstrate explicitly the robot motions so that the robot learns the context (when to react) as well as the motion (how to react).

The applications investigated in this monograph and the roughly described ideas of promising future directions in this chapter reflect just a small portion of possible fields of application. Actually, we see reasonable applications wherever the model can be hierarchically decomposed into parts. The proposed methods are especially suited if the model has to represent a large set of different instances, poses or configurations, or if it has to be applied in real-time. For all these applications this monograph provides important basic concepts, efficient methods as well as practical "how-to" examples.

Bibliography

[1] Aggarwal, J., Ryoo, M.: Human activity analysis: A review. ACM Comput. Surv. 43(3), 16:1–16:43 (2011)

[2] Agrawal, R., Srikant, R.: Fast algorithms for mining association rules in large databases. In: Proceedings of the 20th International Conf. on Very Large Data Bases, San Francisco, CA, USA, pp. 487–499 (1994)

[3] Aha, D.W., Kibler, D., Albert, M.K.: Instance-based learning algorithms. Mach. Learn. 6(1), 37–66 (1991)

[4] Ali, S., Shah, M.: Human action recognition in videos using kinematic features and multiple instance learning. IEEE Transactions on Pattern Analysis and Machine Intelligence 32(2), 288–303 (2010)

[5] Alvarez, J.M., Gevers, T., Lopez, A.: 3d scene priors for road detection. In: Proceedings of the IEEE Conf. on Computer Vision and Pattern Recognition, pp. 57–64 (2010)

[6] Aly, M.: Real Time Detection of Lane Markers in Urban Streets. In: Proceedings of IEEE Intelligent Vehicles Symposium (June 2008)

[7] Arulampalam, M.S., Maskell, S., Gordon, N.: A tutorial on particle filters for online nonlinear/non-gaussian bayesian tracking. IEEE Transactions on Signal Processing 50, 174–188 (2002)

[8] Babich, G.A., Camps, O.I.: Weighted parzen windows for pattern classification. IEEE Transactions on Pattern Analysis and Machine Intelligence 18(5), 567–570 (1996)

[9] Baeza-Yates, R.A., Ribeiro-Neto, B.: Modern Information Retrieval. Addison-Wesley Longman Publishing Co., Inc., Boston (1999)

[10] Barber, D.: Bayesian Reasoning and Machine Learning. Cambridge University Press (2012)

[11] Barlow, H.: What is the computational goal of the neocortex? In: Large Scale Neuronal Theories of the Brain, pp. 1–22. MIT Press, Cambridge (1994)

[12] Bertozzi, M., Broggi, A.: Gold: A parallel real-time stereo vision system for generic obstacle and lane detection. IEEE Transactions on Image Processing 7, 62–81 (1998)

[13] Biederman, I.: Recognition-by-Components: A Theory of Human Image Understanding. Psychological Review 94(2), 115–147 (1987)

[14] Biland, H.-P., Wahl, F.M.: Understanding hough space for polyhedral scene decomposition. Technical report, IBM Research Division (1986)

[15] Bileschi, S.M.: StreetScenes: Towards Scene Understanding in Still Images. PhD thesis, Massachusetts Institute of Technology (2006)

[16] Bishop, C.M.: Pattern Recognition and Machine Learning (Information Science and Statistics), 1st edn. (2007), Corr. 2nd printing edn. (October 2007)

[17] Blei, D.M., Ng, A.Y., Jordan, M.I.: Latent dirichlet allocation. J. Mach. Learn. Res. 3, 993–1022 (2003)

[18] Bobick, A.F., Davis, J.W.: The recognition of human movement using temporal templates. IEEE Transactions on Pattern Analysis and Machine Intelligence 23, 257–267 (2001)

[19] Bouchard, G., Triggs, B.: Hierarchical Part-Based Visual Object Categorization. In: Proceedings of the IEEE Conf. on Computer Vision and Pattern Recognition, San Diego, United States, vol. 1, pp. 710–715 (2005)

[20] Bourdev, L., Maji, S., Brox, T., Malik, J.: Detecting people using mutually consistent poselet activations. In: Daniilidis, K., Maragos, P., Paragios, N. (eds.) ECCV 2010, Part VI. LNCS, vol. 6316, pp. 168–181. Springer, Heidelberg (2010)

[21] Bourdev, L., Malik, J.: Poselets: Body part detectors trained using 3d human pose annotations. In: Proceedings of the IEEE International Conf. on Computer Vision (2009)

[22] Brand, M., Oliver, N., Pentland, A.: Coupled hidden markov models for complex action recognition. Proceedings of the IEEE Conf. on Computer Vision and Pattern Recognition, p. 994 (1997)

[23] Broggi, A., Caraffi, C., Fedriga, R.I., Grisleri, P.: Obstacle detection with stereo vision for off-road vehicle navigation. In: Proceedings of the International IEEE Workshop on Machine Vision for Intelligent Vehicles, p. 65 (2005)

[24] Broggi, A., Cerri, P., Medici, P., Porta, P.P., Ghisio, G.: Real Time Road Signs Recognition. In: Proceedings of IEEE Intelligent Vehicles Symposium, Istanbul, Turkey, pp. 981–986 (2007)

[25] Brooks, L.: Nonanalytic concept formation and memory for instances. In: Cognition and Categorization, pp. 169–211 (1978)

[26] Brostow, G.J., Shotton, J., Fauqueur, J., Cipolla, R.: Segmentation and recognition using structure from motion point clouds. In: Forsyth, D., Torr, P., Zisserman, A. (eds.) ECCV 2008, Part I. LNCS, vol. 5302, pp. 44–57. Springer, Heidelberg (2008)

[27] Burl, M.C., Weber, M., Perona, P.: A probabilistic approach to object recognition using local photometry and global geometry. In: Burkhardt, H., Neumann, B. (eds.) ECCV 1998. LNCS, vol. 1407, pp. 628–641. Springer, Heidelberg (1998)

[28] Burow, M., Wahl, F.M.: Eine verbesserte Version des Kantendetektionsverfahrens nach Mero/Vassy. In: Angewandte Szenenanalyse. Informatik Fachberichte 20, DAGM-Symposium, Berlin, Heidelberg, New York, pp. 36–42 (1979)

[29] Canny, F.J.: A Computational Approach to Edge Detection. IEEE Transactions on Pattern Analysis and Machine Intelligence 8(6), 679–698 (1986)

[30] Chen, Y., Zhu, L., Lin, C., Yuille, A., Zhang, H.: Rapid inference on a novel and/or graph for object detection, segmentation and parsing. In: Advances in Neural Information Processing Systems 20, pp. 289–296. MIT Press, Cambridge (2008)

[31] Comaniciu, D., Meer, P., Member, S.: Mean shift: A robust approach toward feature space analysis. IEEE Transactions on Pattern Analysis and Machine Intelligence 24, 603–619 (2002)

[32] Connolly, C.: The relationship between colour metrics and the appearance of three-dimensional coloured objects. In: Color Research and Applications, vol. 21, pp. 331–337 (1996)

[33] Cover, T., Thomas, J.: Elements of information theory. Wiley, NY (1991)

[34] Crandall, D., Felzenszwalb, P., Huttenlocher, D.: Spatial priors for part-based recognition using statistical models. In: Proceedings of the IEEE Conf. on Computer Vision and Pattern Recognition, Washington, DC, USA, vol. 1, pp. 10–17 (2005)

[35] Csurka, G., Dance, C.R., Fan, L., Willamowski, J., Bray, C.: Visual categorization with bags of keypoints. In: Proceedings of the European Conf. on Computer Vision, Workshop on Statistical Learning in Computer Vision, pp. 1–22 (2004)

[36] Dalal, N.: Finding people in images and videos. PhD thesis, Institut National Polytechnique de Grenoble (July 2006)

[37] Daugman, J.G.: Uncertainty relation for resolution in space, spatial frequency, and orientation optimized by two-dimensional visual cortical filters. J. Opt. Soc. Am. A 2(7), 1160–1169 (1985)

[38] Davis, J.W., Gao, H.: Gender recognition from walking movements using adaptive three-mode pca. In: IEEE Computer Society Conf. on Computer Vision and Pattern Recognition Workshops, Los Alamitos, CA, USA, vol. 1, p. 9 (2004)

[39] Dechter, R., Mateescu, R.: And/or search spaces for graphical models. Artif. Intell. 171(2-3), 73–106 (2007)

[40] Delage, E., Lee, H., Ng, A.Y.: A dynamic bayesian network model for autonomous 3d reconstruction from a single indoor image. In: Proceedings of the IEEE Conf. on Computer Vision and Pattern Recognition, Washington, DC, USA, vol. 2, pp. 2418–2428 (2006)

[41] Detry, R., Pugeault, N., Piater, J.: A probabilistic framework for 3D visual object representation. IEEE Transactions on Pattern Analysis and Machine Intelligence 31(10), 1790–1803 (2009)

[42] Dollar, P., Rabaud, V., Cottrell, G., Belongie, S.: Behavior recognition via sparse spatio-temporal features. In: 2nd Joint IEEE International Workshop on Visual Surveillance and Performance Evaluation of Tracking and Surveillance, pp. 65–72 (2005)

[43] Dollár, P., Wojek, C., Schiele, B., Perona, P.: Pedestrian detection: An evaluation of the state of the art. IEEE Transactions on Pattern Analysis and Machine Intelligence 34(4) (2012)

[44] Dorkó, G., Dorkó, G., Schmid, C., Schmid, C., Lear, P.: Object class recognition using discriminative local features. IEEE Transactions on Pattern Analysis and Machine Intelligence (2005)

[45] Doucet, A., Godsill, S., Andrieu, C.: On sequential monte carlo sampling methods for bayesian filtering. Statistics and Computing 10(3), 197–208 (2000)

[46] Efros, A.A., Berg, A.C., Mori, G., Malik, J.: Recognizing action at a distance. In: Proceedings of the IEEE International Conf. on Computer Vision, Nice, France, pp. 726–733 (2003)

[47] Engelbrecht, J.R., Wahl, F.M.: Polyhedral object recognition using hough-space features. Pattern Recognition 21, 155–167 (1988)

[48] Epshtein, B., Ullman, S.: Feature hierarchies for object classification. In: Proceedings of the IEEE International Conf. on Computer Vision, Washington, DC, USA, pp. 220–227 (2005)

[49] Epshtein, B., Ullman, S.: Semantic hierarchies for recognizing objects and parts. In: Proceedings of the IEEE Conf. on Computer Vision and Pattern Recognition, pp. 1–8 (2007)

[50] Ess, A., Leibe, B., Schindler, K., Van Gool, L.: Moving obstacle detection in highly dynamic scenes. In: Proceedings of the 2009 IEEE international Conf. on Robotics and Automation, Piscataway, NJ, USA, pp. 4451–4458 (2009)

[51] Ess, A., Mueller, T., Grabner, H., Gool, L.J.V.: Segmentation-based urban traffic scene understanding. In: Proceedings of the British Machine Vision Conf., London, UK (September 2009)

[52] Ester, M., Kriegel, H.-P., Sander, J., Xu, X.: A density-based algorithm for discovering clusters in large spatial databases with noise. In: Second International Conf. on Knowledge Discovery and Data Mining, Portland, OR, pp. 226–231. AAAI Press (1996)

[53] Ettinger, G.J.: Hierarchical object recognition using libraries of parameterized model sub-parts. Technical report, Massachusetts Institute of Technology, Cambridge, MA, USA (1987)

[54] Ettinger, G.J.: Large Hierarchical Object Recognition Using Libraries of Parameterized Model Sub-parts. In: Proceedings of the IEEE Conf. on Computer Vision and Pattern Recognition, Ann Arbor, Michigan (June 1988)

[55] Fei-Fei, L., Li, L.-J.: What, Where and Who? Telling the Story of an Image by Activity Classification, Scene Recognition and Object Categorization. In: Cipolla, R., Battiato, S., Farinella, G.M. (eds.) Computer Vision. SCI, vol. 285, pp. 157–171. Springer, Heidelberg (2010)

[56] Felzenszwalb, P.F., Girshick, R.B., McAllester, D., Ramanan, D.: Object detection with discriminatively trained part-based models. IEEE Transactions on Pattern Analysis and Machine Intelligence 32(9), 1627–1645 (2010)

[57] Felzenszwalb, P.F., Huttenlocher, D.P.: Pictorial structures for object recognition. Int. J. Comput. Vision 61(1), 55–79 (2005)

[58] Fergus, R., Perona, P., Zisserman, A.: Object class recognition by unsupervised scale-invariant learning. In: Proceedings of the IEEE Conf. on Computer Vision and Pattern Recognition, vol. 2, pp. 264–271 (June 2003)

[59] Fergus, R., Perona, P., Zisserman, A.: A sparse object category model for efficient learning and exhaustive recognition. In: Proceedings of the IEEE Conf. on Computer Vision and Pattern Recognition, pp. 380–387 (2005)

[60] Fergus, R., Perona, P., Zisserman, A.: Weakly supervised scale-invariant learning of models for visual recognition. Int. J. Comput. Vision 71(3), 273–303 (2007)

[61] Ferrari, V., Fevrier, L., Jurie, F., Schmid, C.: Groups of adjacent contour segments for object detection. IEEE Transactions on Pattern Analysis and Machine Intelligence 30(1), 36–51 (2008)

[62] Ferrari, V., Jurie, F., Schmid, C.: Accurate object detection with deformable shape models learnt from images. In: Proceedings of the IEEE Conf. on Computer Vision and Pattern Recognition (2007)

[63] Fidler, S.: Recognizing visual object categories with subspace methods and a learned hierarchical shape vocabulary. PhD thesis, University of Ljubljana, Faculty of computer and information science (2010)

[64] Fidler, S., Boben, M., Leonardis, A.: Similarity-based cross-layered hierarchical representation for object categorization. In: Proceedings of the IEEE Conf. on Computer Vision and Pattern Recognition (2008)

[65] Fidler, S., Boben, M., Leonardis, A.: A coarse-to-fine taxonomy of constellations for fast multi-class object detection. In: Daniilidis, K., Maragos, P., Paragios, N. (eds.) ECCV 2010, Part V. LNCS, vol. 6315, pp. 687–700. Springer, Heidelberg (2010)

[66] Fidler, S., Leonardis, A.: Towards scalable representations of object categories: Learning a hierarchy of parts. In: Proceedings of the IEEE Conf. on Computer Vision and Pattern Recognition, Minnesota, USA (June 2007)

[67] Fine, S.: The hierarchical hidden markov model: Analysis and applications. In: Machine Learning, pp. 41–62 (1998)

[68] Fischler, M., Elschlager, R.: The representation and matching of pictorial structures. IEEE Transactions on Computers 22(1), 67–92 (1973)

[69] Fiser, J., Aslin, R.N.: Encoding multielement scenes: Statistical learning of visual feature hierarchies. Journal of Experimental Psychology: General, 521–537 (2005)

[70] Fiser, J., Aslin, R.N.: Unsupervised Statistical Learning of Higher-Order Spatial Structures From Visual Scenes. Psychological Science (November 12, 2001)

[71] Fleuret, F., Geman, D.: Coarse-to-fine face detection. International Journal of Computer Vision (IJCV) 41(1/2), 85–107 (2001)

[72] Frank, D.: Road markings recognition. In: International Conf. on Image Processing, Lausanne, Switzerland, vol. 2, pp. 669–672 (September 1996)

[73] Fukushima, K., Miyake, S., Ito, T.: Neocognitron: A neural network model for a mechanism of visual pattern recognition. IEEE Transactions on Systems, Man, and Cybernetics SMC-13, 826–834 (1983)

[74] Gavrila, D.: Traffic sign recognition revisited. In: Mustererkennung 1999. DAGM-Symposium, London, UK., vol. 21, pp. 86–93 (1999)

[75] Gavrila, D.M.: Pedestrian detection from a moving vehicle. In: Vernon, D. (ed.) ECCV 2000. LNCS, vol. 1843, pp. 37–49. Springer, Heidelberg (2000)

[76] Gavrila, D.M.: Multi-feature hierarchical template matching using distance transforms. In: Proceedings of the 14th International Conf. on Pattern Recognition, Washington, DC, USA, vol. 1, p. 439 (1998)

[77] Gavrila, D.M., Munder, S.: Multi-cue pedestrian detection and tracking from a moving vehicle. International Journal of Computer Vision 73, 41–59 (2007)

[78] Geiger, A., Lauer, M., Urtasun, R.: A generative model for 3d urban scene understanding from movable platforms. In: Proceedings of the IEEE Conf. on Computer Vision and Pattern Recognition, Colorado Springs, USA (June 2011)

[79] Geismann, P., Schneider, G.: A two-staged approach to vision-based pedestrian recognition using haar and hog features. In: Proceedings of IEEE Intelligent Vehicles Symposium (2008)

[80] Geman, S., Potter, D., Chi, Z.: Composition systems. Quarterly of Applied Mathematics 60(4), 707–736 (2002)

[81] Gorelick, L., Blank, M., Shechtman, E., Irani, M., Basri, R.: Actions as space-time shapes. IEEE Transactions on Pattern Analysis and Machine Intelligence 29(12), 2247–2253 (2007)

[82] Gould, R.J.: Graph Theory. Benjamin-Cummings Publishing Company, Subs of Addison Wesley Longman, Inc. (1988)

[83] Griffin, G., Holub, A., Perona, P.: Caltech-256 object category dataset. Technical Report 7694, California Institute of Technology (2007)

[84] Gupta, A., Davis, L.S.: Objects in action: An approach for combining action understanding and object perception. In: Proceedings of the IEEE Conf. on Computer Vision and Pattern Recognition, pp. 1–8 (2007)

[85] Gupta, A., Efros, A.A., Hebert, M.: Blocks world revisited: Image understanding using qualitative geometry and mechanics. In: Daniilidis, K., Maragos, P., Paragios, N. (eds.) ECCV 2010, Part IV. LNCS, vol. 6314, pp. 482–496. Springer, Heidelberg (2010)

[86] Harris, C., Stephens, M.: A combined corner and edge detector. In: Proceedings of the 4th Alvey Vision Conf., pp. 147–151 (1988)

[87] Hartley, R.I., Zisserman, A.: Multiple View Geometry in Computer Vision, 2nd edn. Cambridge University Press (2004)

[88] Hastings, W.K.: Monte carlo sampling methods using markov chains and their applications. Biometrika 57(1), 97–109 (1970)

[89] Hogg, D.: Model-based vision: A program to see a walking person. Image and Vision Computing 1(1), 5–20 (1983)

[90] Hoiem, D., Efros, A.A., Hebert, M.: Recovering surface layout from an image. Int. J. Comput. Vision 75(1), 151–172 (2007)

[91] Hoiem, D., Efros, A.A., Hebert, M.: Closing the loop on scene interpretation. In: Proceedings of the IEEE Conf. on Computer Vision and Pattern Recognition (June 2008)

[92] Hospedales, T.M., Gong, S., Xiang, T.: A markov clustering topic model for mining behaviour in video. In: IEEE 12th International Conf. on Computer Vision, Kyoto, Japan, pp. 1165–1172 (September 2009)

[93] Hospedales, T.M., Li, J., Gong, S., Xiang, T.: Identifying rare and subtle behaviors: A weakly supervised joint topic model. IEEE Transactions on Pattern Analysis and Machine Intelligence 33, 2451–2464 (2011)

[94] Hua, G., Yang, M.-H., Wu, Y.: Learning to estimate human pose with data driven belief propagation. In: Proceedings of the IEEE Conf. on Computer Vision and Pattern Recognition, Washington, DC, USA, vol. 2, pp. 747–754 (2005)

[95] Hubel, D., Wiesel, T.: Receptive fields, binocular interaction and functional architecture in the cat's visual cortex. J. Phys. 160, 106–154 (1962)

[96] Hubel, D., Wiesel, T.: Receptive fields and functional architecture in two nonstriate visual areas of the cat. J. Neurophys. 28, 229–289 (1965)

[97] Hueckel, M.H.: An operator which locates edges in digitized pictures. Journal of the ACM 18, 113–125 (1977)

[98] Ihler, A.T., Sudderth, E., Freeman, W., Willsky, A.: Efficient multiscale sampling from products of gaussian mixtures. In: Advances in Neural Information Processing Systems 17, p. 2003. MIT Press (2003)

[99] Ioffe, S., Forsyth, D.A.: Probabilistic methods for finding people. Int. J. Comput. Vision 43(1), 45–68 (2001)

[100] Isard, M.: Pampas: Real-valued graphical models for computer vision. In: Proceedings of the IEEE Conf. on Computer Vision and Pattern Recognition, pp. 613–620 (2003)

[101] Isard, M., Blake, A.: Condensation - conditional density propagation for visual tracking. International Journal of Computer Vision 29, 5–28 (1998)

[102] Jeon, B., Landgrebe, D.: Fast parzen density estimation using clustering-based branch andbound. IEEE Transactions on Pattern Analysis and Machine Intelligence 16(9), 950–954 (1994)

[103] Jiang, Z., Lin, Z., Davis, L.S.: Recognizing human actions by learning and matching shape-motion prototype trees. IEEE Transactions on Pattern Analysis and Machine Intelligence 34(3), 533–547 (2012)

[104] Jin, Y., Geman, S.: Context and hierarchy in a probabilistic image model. In: Proceedings of the IEEE Conf. on Computer Vision and Pattern Recognition, Washington, DC, USA, vol. 2, pp. 2145–2152 (2006)

[105] Ju, S.X., Black, M.J., Yacoob, Y.: Cardboard people: A parameterized model of articulated image motion. In: 2nd International Conf. on Automatic Face and Gesture Recognition, Killington, Vermont, USA, October 14-16, pp. 38–44 (1996)

[106] Jung, H.G.: Target Position Designation Methods for Parking Assistant System. PhD thesis, Yonsei University, School of Electrical and Electronic Engineering (August 2008)

[107] Jung, H.G., Kim, D.S., Yoon, P.J., Kim, J.: Parking slot markings recognition for automatic parking assist system. In: Proceedings of IEEE Intelligent Vehicles Symposium, pp. 106–113 (2006)

[108] Kale, A., Rajagopalan, A., Cuntoor, N., Krueger, V.: Human identification using gait. In: International Conf. on Automatic Face and Gesture Recognition, Washington DC, USA (2002)

[109] Ke, Y., Sukthankar, R., Hebert, M.: Event detection in crowded videos. In: Proceedings of the IEEE International Conf. on Computer Vision, pp. 1–8 (2007)

[110] Kim, Z.: Robust lane detection and tracking in challenging scenarios. Trans. Intell. Transport. Sys. 9(1), 16–26 (2008)

[111] Kluge, K., Thorpe, C.: The yarf system for vision-based road following. Mathematical and Computer Modelling 22(4-7), 213–233 (1995)

[112] Koller, D., Friedman, N.: Probabilistic Graphical Models: Principles and Techniques. MIT Press (2009)

[113] Koseck'a, C.J.T.J., Blasi, R., Malik, J.: A comparative study of vision-based lateral control strategies forautonomous highway driving. International Journal of Robotics Research 18, 442–453 (1999)

[114] Krempp, S., Geman, D., Amit, Y.: Sequential learning of reusable parts for object detection. Technical report, CS Johns Hopkins (2002)

[115] Kschischang, F., Member, S., Frey, B.J., Andrea Loeliger, H.: Factor graphs and the sum-product algorithm. IEEE Transactions on Information Theory 47, 498–519 (2001)

[116] Kuettel, D., Breitenstein, M.D., Gool, L.V., Ferrari, V.: What's going on? discovering spatio-temporal dependencies in dynamic scenes. In: Proceedings of the IEEE Conf. on Computer Vision and Pattern Recognition (June 2010)

[117] Lades, M., Vorbrüggen, J.C., Buhmann, J., Lange, J., Malsburg, C.V.D., Würtz, R.P., Konen, W.: Distortion invariant object recognition in the dynamic link architecture. IEEE Trans. Computers 42, 300–311 (1993)

[118] Laptev, I., Lindeberg, T.: Space-time interest points. In: Proceedings of the IEEE International Conf. on Computer Vision, pp. 432–439 (2003)

[119] Lauritzen, S.L.: Graphical Models. Oxford University Press, Oxford (1996)

[120] Lauritzen, S.L., Spiegelhalter, D.J.: Local computations with probabilities on graphical structures and their application to expert systems. Journal of the Royal Statistical Society, Series B (50(2)), 157–224 (1988)

[121] LeCun, Y., Bengio, Y.: Convolutional networks for images, speech, and time-series. In: Arbib, M.A. (ed.) The Handbook of Brain Theory and Neural Networks. MIT Press (1995)

[122] LeCun, Y., Bottou, L., Bengio, Y., Haffner, P.: Gradient-based learning applied to document recognition. Proceedings of the IEEE 86(11), 2278–2324 (1998)

[123] Lee, D.C., Hebert, M., Kanade, T.: Geometric reasoning for single image structure recovery. In: Proceedings of the IEEE Conf. on Computer Vision and Pattern Recognition (June 2009)

[124] Lee, M.W., Nevatia, R.: Human pose tracking in monocular sequence using multi-level structured models. IEEE Transactions on Pattern Analysis and Machine Intelligence 31(1), 27–38 (2009)

[125] Leibe, B., Leonardis, A., Schiele, B.: Combined object categorization and segmentation with an implicit shape model. In: Proceedings of the European Conf. on Computer Vision, Workshop on Statistical Learning in Computer Vision, pp. 17–32 (2004)

[126] Leibe, B., Schiele, B.: Interleaved object categorization and segmentation. In: Proceedings of the British Machine Vision Conf., pp. 759–768 (2003)

[127] Leibe, B., Schiele, B.: Scale-invariant object categorization using a scale-adaptive mean-shift search. In: Rasmussen, C.E., Bülthoff, H.H., Schölkopf, B., Giese, M.A. (eds.) DAGM 2004. LNCS, vol. 3175, pp. 145–153. Springer, Heidelberg (2004)

[128] Lepetit, V., Lagger, P., Fua, P.: Randomized trees for real-time keypoint recognition. In: Proceedings of the IEEE Conf. on Computer Vision and Pattern Recognition, pp. 775–781 (2005)

[129] Li, F.-F., Perona, P.: A bayesian hierarchical model for learning natural scene categories. In: Proceedings of the IEEE Conf. on Computer Vision and Pattern Recognition, Washington, DC, USA, vol. 2, pp. 524–531 (2005)

[130] Lin, Z., Davis, L.S., Doermann, D.S., DeMenthon, D.: Hierarchical part-template matching for human detection and segmentation. In: Proceedings of the IEEE International Conf. on Computer Vision, pp. 1–8 (2007)

[131] Lindeberg, T.: Scale-Space Theory in Computer Vision. Kluwer Academic Publishers, Norwell (1994)

[132] Lindeberg, T.: Scale-space: A framework for handling image structures at multiple scales. In: CERN School of Computing, Egmond aan Zee, The Netherlands, September 8-21 (1996)

[133] Lindeberg, T.: Feature detection with automatic scale selection. International Journal of Computer Vision 30, 79–116 (1998)

[134] Lowe, D.G.: Object recognition from local scale-invariant features. In: Proceedings of the IEEE International Conf. on Computer Vision, Washington, DC, USA, p. 1150 (1999)

[135] Lowe, D.G.: Distinctive image features from scale-invariant keypoints. Int. J. Comput. Vision 60(2), 91–110 (2004)

[136] Lu, C., Ferrier, N.J.: Repetitive motion analysis: Segmentation and event classification. IEEE Transactions on Pattern Analysis and Machine Intelligence 26(2), 258–263 (2004)

[137] Lucas, B.D., Kanade, T.: An iterative image registration technique with an application to stereo vision (ijcai). In: Proceedings of the 7th International Joint Conf. on Artificial Intelligence, pp. 674–679 (April 1981)

[138] MacKay, D.J.C.: Introduction to Monte Carlo methods. In: Learning in Graphical Models. NATO Science Series, pp. 175–204. Kluwer Academic Press (1998)

[139] Mallot, H.A., Bülthoff, H.H., Little, J.J., Bohrer, S.: Inverse perspective mapping simplifies optical flow computation and obstacle detection. Biological Cybernetics 64(3), 177–185 (1991)

[140] Mardia, K.V., Jupp, P.E.: Directional Statistics. John Wiley and Sons, New York (2000)

[141] Marée, R., Geurts, P., Piater, J.H., Wehenkel, L.: Random subwindows for robust image classification. In: Proceedings of the IEEE Conf. on Computer Vision and Pattern Recognition, pp. 34–40 (2005)

[142] Marr, D., Nishihara, H.K.: Representation and recognition of the spatial organisation of three-dimensional shapes. Proceedings of the Royal Society of London B200, 269–294 (1978)

[143] Marszałek, M., Laptev, I., Schmid, C.: Actions in context. In: Proceedings of the IEEE Conf. on Computer Vision and Pattern Recognition (2009)

[144] Masquelier, T., Thorpe, S.J.: Unsupervised Learning of Visual Features through Spike Timing Dependent Plasticity. PLoS Comput Biol. 3(2), e31+ (2007)

[145] McCall, J.C., Trivedi, M.M.: Video-based lane estimation and tracking for driver assistance: survey, system, and evaluation. Trans. Intell. Transport. Sys. 7(1), 20–37 (2006)

[146] Mero, L., Vassy, Z.: A simplified and fast version of the hueckel operator for finding optimal edges in pictures. In: Proc. 4th Int. Conf. on Artificial Intelligence, pp. 650–655 (1975)

[147] Micilotta, A., Ong, E., Bowden, R.: Detection and tracking of humans by probabilistic body part assembly. In: Proceedings of the British Machine Vision Conf., Oxford, UK (2005)

[148] Mikolajczyk, K.: Multiple object class detection with a generative model. In: Proceedings of the IEEE Conf. on Computer Vision and Pattern Recognition, pp. 26–36 (2006)

[149] Moeslund, T.B., Granum, E.: A survey of computer vision-based human motion capture. Computer Vision and Image Understanding 81(3), 231–268 (2001)

[150] Moeslund, T.B., Hilton, A., Krüger, V.: A survey of advances in vision-based human motion capture and analysis. Computer Vision and Image Understanding 104(2), 90–126 (2006)

[151] Moore, D., Essa, I., Hayes, M.: Exploiting Human Actions and Object Context for Recognition Tasks. In: Proceedings of IEEE International Conf. on Computer Vision, Corfu, Greece (March 1999)

[152] Moravec, H.: Towards automatic visual obstacle avoidance. In: Proceedings of the 5th International Joint Conf. on Artificial Intelligence, p. 584 (August 1977)

[153] Moravec, H., Elfes, A.E.: High resolution maps from wide angle sonar. In: Proceedings of the 1985 IEEE International Conf. on Robotics and Automation, pp. 116–121 (March 1985)

[154] Mori, G.: Guiding model search using segmentation. In: Proceedings of the IEEE International Conf. on Computer Vision, Washington, DC, USA, pp. 1417–1423 (2005)

[155] Mori, G., Malik, J.: Estimating human body configurations using shape context matching. In: Heyden, A., Sparr, G., Nielsen, M., Johansen, P. (eds.) ECCV 2002, Part III. LNCS, vol. 2352, pp. 666–680. Springer, Heidelberg (2002)

[156] Mori, G., Ren, X., Efros, A., Malik, J.: Recovering human body configurations: Combining segmentation and recognition. In: Proceedings of the IEEE Conf. on Computer Vision and Pattern Recognition, vol. 2, pp. 326–333 (2004)

[157] Morita, K., Atlam, E.-S., Fuketra, M., Tsuda, K., Oono, M.: J.-i. Aoe. Word classification and hierarchy using co-occurrence word information. Inf. Process. Manage. 40(6), 957–972 (2004)

[158] Mouragnon, E., Lhuillier, M., Dhome, M., Dekeyser, F., Sayd, P.: Monocular vision based slam for mobile robots. In: International Conf. on Pattern Recognition, vol. 3, pp. 1027–1031 (2006)

[159] Murphy, K., Weiss, Y., Jordan, M.: Loopy belief propagation for approximate inference:an empirical study. Uncertainty in Artificial Intelligence (1999)

[160] Murthy, S.K.: Automatic construction of decision trees from data: A multi-disciplinary survey. Data Mining and Knowledge Discovery 2, 345–389 (1997)

[161] Musso, C., Oudjane, N., Gland, F.L.: Improving regularised particle filters. In: Sequential Monte Carlo Methods in Practice, pp. 247–271 (2001)

[162] Mutch, J., Lowe, D.G.: Multiclass object recognition with sparse, localized features. In: Proceedings of the IEEE Conf. on Computer Vision and Pattern Recognition, New York, pp. 11–18 (June 2006)

[163] Natarajan, P., Nevatia, R.: View and scale invariant action recognition using multiview shape-flow models. In: Proceedings of the IEEE Conf. on Computer Vision and Pattern Recognition (2008)

[164] Nedovic, V., Smeulders, A.W.M., Redert, A., Geusebroek, J.-M.: Stages as models of scene geometry. IEEE Transactions on Pattern Analysis and Machine Intelligence, 1673–1687 (2010)

[165] Nene, S.A., Nayar, S.K., Murase, H.: Columbia object image library (coil-100). Technical Report CUCS-006-96, California Institute of Technology (February 1996)

[166] Niebles, J., Fei-Fei, L.: A hierarchical model of shape and appearance for human action classification. In: Proceedings of the IEEE Conf. on Computer Vision and Pattern Recognition (2007)

[167] Niebles, J.C., Wang, H., Fei-Fei, L.: Unsupervised learning of human action categories using spatial-temporal words. Int. J. Comput. Vision 79(3), 299–318 (2008)

[168] Nieto, M., Salgado, L., Jaureguizar, F.: Robust road modeling based on a hierarchical bipartite graph. In: Proceedings of IEEE Intelligent Vehicles Symposium (2008)

[169] Nister, D., Stewenius, H.: Scalable recognition with a vocabulary tree. In: Proceedings of the IEEE Conf. on Computer Vision and Pattern Recognition, Washington, DC, USA, vol. 2, pp. 2161–2168 (2006)

[170] Nock, R., Nielsen, F.: Statistical region merging. IEEE Transactions on Pattern Analysis and Machine Intelligence 26(11), 1452–1458 (2004)

[171] Obdrzálek, S., Matas, J.: Sub-linear indexing for large scale object recognition. In: Proceedings of the British Machine Vision Conf. (2005)

[172] Oliva, A.: Gist of the scene. In: Itti, L., Rees, G., Tsotsos, J.K. (eds.) The Encyclopedia of Neurobiology of Attention, pp. 251–256. Elsevier, San Diego (2005)

[173] Oliva, A., Torralba, A.: Building the gist of a scene: the role of global image features in recognition. In: Progress in Brain Research, p. 2006 (2006)

[174] Oliver, N., Horvitz, E., Garg, A.: Layered representations for human activity recognition. In: Proceedings of the 4th IEEE International Conf. on Multimodal Interfaces, Washington, DC, USA, p. 3 (2002)

[175] Ommer, B., Buhmann, J.M.: Object categorization by compositional graphical models. In: Rangarajan, A., Vemuri, B.C., Yuille, A.L. (eds.) EMMCVPR 2005. LNCS, vol. 3757, pp. 235–250. Springer, Heidelberg (2005)

[176] Ommer, B., Buhmann, J.M.: Learning compositional categorization models. In: Leonardis, A., Bischof, H., Pinz, A. (eds.) ECCV 2006. LNCS, vol. 3953, pp. 316–329. Springer, Heidelberg (2006)

[177] Ommer, B., Buhmann, J.M.: Learning the compositional nature of visual object categories for recognition. IEEE Transactions on Pattern Analysis and Machine Intelligence 32(3), 501–516 (2010)

[178] Opelt, A., Pinz, A., Zisserman, A.: Learning an alphabet of shape and appearance for multi-class object detection. Int. J. Comput. Vision 80(1), 16–44 (2008)

[179] Pearl, J.: Probabilistic Reasoning in Intelligent Systems: Networks of PlausibleInference. Morgan Kaufmann (1988)

[180] Peursum, P., West, G., Venkatesh, S.: Combining image regions and human activity for indirect object recognition in indoor wide-angle views. In: Proceedings of the IEEE International Conf. on Computer Vision, Washington, DC, USA, pp. 82–89 (2005)

[181] Philbin, J., Chum, O., Isard, M., Sivic, J., Zisserman, A.: Object retrieval with large vocabularies and fast spatial matching. In: Proceedings of the IEEE Conf. on Computer Vision and Pattern Recognition (2007)

[182] Polana, R., Nelson, A.: Low level recognition of human motion (or how to get your man without finding his body parts. In: IEEE Computer Society Workshop on Motion of Non-Rigid and Articulated Objects, pp. 77–82. Press (1994)

[183] Poppe, R.W.: A survey on vision-based human action recognition. Image and Vision Computing 28(6), 976–990 (2010)

[184] Rabiner, L.R.: A tutorial on hidden markov models and selected applications in speech recognition. In: Waibel, A., Lee, K.-F. (eds.) Readings in Speech Recognition, pp. 267–296. Morgan Kaufmann Publishers Inc., San Francisco (1990)

[185] Ramanan, D.: Learning to parse images of articulated bodies. In: Advances in Neural Information Processing Systems 19, pp. 1129–1136 (2006)

[186] Ramanan, D., Forsyth, D.A., Zisserman, A.: Strike a pose: Tracking people by finding stylized poses. In: Proceedings of the IEEE Conf. on Computer Vision and Pattern Recognition, Washington, DC, USA, vol. 1, pp. 271–278 (2005)

[187] Rao, C., Yilmaz, A., Shah, M.: View-invariant representation and recognition of actions. International Journal of Computer Vision 50(2), 203–226 (2002)

[188] Reed, S.K.: Pattern recognition and categorization. Cognitive Psychology 3(3), 382–407 (1972)

[189] Ren, X., Berg, A.C., Malik, J.: Recovering human body configurations using pairwise constraints between parts. In: Proceedings of the IEEE International Conf. on Computer Vision, Washington, DC, USA, pp. 824–831 (2005)

[190] Reng, L., Moeslund, T.B., Granum, E.: Finding motion primitives in human body gestures. In: Gibet, S., Courty, N., Kamp, J.-F. (eds.) GW 2005. LNCS (LNAI), vol. 3881, pp. 133–144. Springer, Heidelberg (2006)

[191] Riemenschneider, H., Donoser, M., Bischof, H.: Bag of optical flow volumes for image sequence recognition. In: Proceedings of the British Machine Vision Conf. (2009)

[192] Riesenhuber, M., Poggio, T.: Hierarchical models of object recognition in cortex. Nature Neuroscience 2, 1019–1025 (1999)

[193] Roberts, T.J., McKenna, S.J., Ricketts, I.W.: Human pose estimation using learnt probabilistic region similarities and partial configurations. In: Pajdla, T., Matas, J. (eds.) ECCV 2004. LNCS, vol. 3024, pp. 291–303. Springer, Heidelberg (2004)

[194] Robertson, N., Reid, I.: Behaviour understanding in video: a combined method. In: Proceedings of the IEEE International Conf. on Computer Vision, Beijing, Chine, vol. 1, pp. 808–814 (2005)

[195] Rohr, K.: Incremental recognition of pedestrians from imagesequences. In: Proceedings of the IEEE Conf. on Computer Vision and Pattern Recognition, vol. 1, 2 (1993)

[196] Ronfard, R., Schmid, C., Triggs, B.: Learning to parse pictures of people. In: Heyden, A., Sparr, G., Nielsen, M., Johansen, P. (eds.) ECCV 2002, Part IV. LNCS, vol. 2353, pp. 700–714. Springer, Heidelberg (2002)

[197] Rosten, E., Drummond, T.: Fusing points and lines for high performance tracking. In: Proceedings of the IEEE International Conf. on Computer Vision, vol. 2, pp. 1508–1511 (2005)

[198] Rosten, E., Drummond, T.: Machine learning for high-speed corner detection. In: Leonardis, A., Bischof, H., Pinz, A. (eds.) ECCV 2006, Part I. LNCS, vol. 3951, pp. 430–443. Springer, Heidelberg (2006)

[199] Ryoo, M.S., Aggarwal, J.K.: Recognition of composite human activities through context-free grammar based representation. In: Proceedings of the IEEE Conf. on Computer Vision and Pattern Recognition, Washington, DC, USA, vol. 2, pp. 1709–1718 (2006)

[200] Ryoo, M.S., Aggarwal, J.K.: Hierarchical recognition of human activities interacting with objects. In: Proceedings of the IEEE Conf. on Computer Vision and Pattern Recognition (2007)

[201] Sapp, B., Toshev, A., Taskar, B.: Cascaded models for articulated pose estimation. In: Daniilidis, K., Maragos, P., Paragios, N. (eds.) ECCV 2010, Part II. LNCS, vol. 6312, pp. 406–420. Springer, Heidelberg (2010)

[202] Savarese, S., Fei-Fei, L.: 3d generic object categorization, localization and pose estimation. In: Proceedings of the IEEE International Conf. on Computer Vision, Rio de Janeiro, Brazil (October 2007)

[203] Saxena, A., Sun, M., Ng, A.Y.: Make3d: Learning 3d scene structure from a single still image. IEEE Transactions on Pattern Analysis and Machine Intelligence 31(5), 824–840 (2009)

[204] Scalzo, F.: Learning Visual Feature Hierarchies. PhD thesis, University of Liege, Belgium (2007-2008)

[205] Scalzo, F., Piater, J.H.: Statistical learning of visual feature hierarchies. In: Proceedings of the IEEE Conf. on Computer Vision and Pattern Recognition, vol. 3, p. 44 (2005)

[206] Scharstein, D., Szeliski, R.: A taxonomy and evaluation of dense two-frame stereo correspondence algorithms. Int. J. Comput. Vision 47(1-3), 7–42 (2002)

[207] Schuldt, C., Laptev, I., Caputo, B.: Recognizing human actions: a local svm approach. In: Proceedings of the 17th International Conf. on Pattern Recognition, vol. 3, pp. 32–36 (2004)

[208] Sebsadji, Y., Tarel, J.-P., Foucher, P., Charbonnier, P.: Robust road marking extraction in urban environments using stereo images. In: Proceedings of IEEE Intelligent Vehicles Symposium, San Diego, California, USA, pp. 394–400 (2010)

[209] Sermanet, P., LeCun, Y.: Traffic sign recognition with multi-scale convolutional networks. In: The 2011 International Joint Conf. on Neural Networks, San Jose, California, USA, pp. 2809–2813 (2011)

[210] Serre, T., Kouh, M., Cadieu, C., Knoblich, U., Kreiman, G., Poggio, T.: A theory of object recognition: Computations and circuits in the feedforward path of the ventral stream in primate visual cortex. In: AI Memo (2005)

[211] Serre, T., Wolf, L., Bileschi, S., Riesenhuber, M., Poggio, T.: Robust object recognition with cortex-like mechanisms. IEEE Transactions on Pattern Analysis and Machine Intelligence 29, 411–426 (2007)

[212] Shakhnarovich, G., Viola, P., Darrell, T.: Fast pose estimation with parameter sensitive hashing. In: Proceedings of the IEEE International Conf. on Computer Vision, pp. 750–757 (2003)

[213] Shechtman, E., Irani, M.: Space-time behavior based correlation –or– how to tell if two underlying motion fields are similar without computing them? IEEE Transactions on Pattern Analysis and Machine Intelligence 29(11), 2045–2056 (2007)

[214] Sigal, L., Bhatia, S., Roth, S., Black, M.J., Isard, M.: Tracking loose-limbed people. In: Proceedings of the IEEE Conf. on Computer Vision and Pattern Recognition, Los Alamitos, CA, USA, vol. 1, pp. 421–428 (2004)

[215] Silverman, B.W.: Density Estimation for Statistics and Data Analysis. Chapman & Hall/CRC (April 1986)

[216] Sivic, J., Zisserman, A.: Video Google: A text retrieval approach to object matching in videos. In: Proceedings of the International Conf. on Computer Vision, vol. 2, pp. 1470–1477 (October 2003)

[217] Sminchisescu, C., Kanaujia, A., Metaxas, D.: Conditional models for contextual human motion recognition. Comput. Vis. Image Underst. 104(2), 210–220 (2006)

[218] Song, G.Y., Lee, K.Y., Lee, J.W.: Vehicle detection by edge-based candidate generation and appearance-based classification. In: Proceedings of IEEE Intelligent Vehicles Symposium (2008)

[219] Spehr, J., Gietzelt, M., Wegel, S., Költzsch, Y., Winkelbach, S., Marschollek, M., Gövercin, M., Wahl, F.M., Haux, R., Steinhagen-Thiessen, E.: Vermessung von gangparametern zur sturzprädikation durch vision- und beschleunigungssensorik. In: 4. Deutscher AAL-Kongress, Berlin, Germany (January 2011)

[220] Spehr, J., Gövercin, M., Winkelbach, S., Steinhagen-Thiessen, E., Wahl, F.M.: Visual fall detection in home environments. In: 6th Int. Conf. of the Int. Soc. for Gerontechnology, Pisa, Italy (June 2008)

[221] Spehr, J., Islami, M., Winkelbach, S., Wahl, F.M.: Recognition of human behavior patterns using depth information and gaussian feature maps. In: GI/GMDS-Workshop, Braunschweig (2012)

[222] Spehr, J., Rosebrock, D., Mossau, D., Auer, R., Brosig, S., Wahl, F.M.: Hierarchical scene understanding for intelligent vehicles. In: Proceedings of IEEE Intelligent Vehicles Symposium, Baden-Baden, Germany, pp. 1142–1147 (June 2011)

[223] Spehr, J., Winkelbach, S., Wahl, F.M.: Hierarchical pose estimation for human gait analysis. Comput. Methods Prog. Biomed. 106(2), 104–113 (2012)

[224] Srinivasan, P., Shi, J.: Bottom-up recognition and parsing of the human body. In: Yuille, A.L., Zhu, S.-C., Cremers, D., Wang, Y. (eds.) EMMCVPR 2007. LNCS, vol. 4679, pp. 153–168. Springer, Heidelberg (2007)

[225] Sturgess, P., Alahari, K., Ladicky, L., Torr, P.H.S.: Combining appearance and structure from motion features for road scene understanding. In: Proceedings of the British Machine Vision Conf. (2009)

[226] Sudderth, E.B.: Graphical Models for Visual Object Recognition and Tracking. PhD thesis, Massachusetts Institute of Technology (2006)

[227] Sudderth, E.B., Ihler, E.T., Freeman, W.T., Willsky, A.: Nonparametric belief propagation. In: Proceedings of the IEEE Conf. on Computer Vision and Pattern Recognition, pp. 605–612 (2003)

[228] Sudderth, E.B., Torralba, A., Freeman, W.T., Willsky, A.S.: Learning hierarchical models of scenes, objects, and parts. In: IEEE Intl. Conf. on Computer Vision, pp. 1331–1338 (2005)

[229] Sullivan, J., Carlsson, S.: Recognizing and tracking human action. In: Heyden, A., Sparr, G., Nielsen, M., Johansen, P. (eds.) ECCV 2002, Part I. LNCS, vol. 2350, pp. 629–644. Springer, Heidelberg (2002)

[230] Tarr, M.J., Williams, P., Hayward, W.G., Gauthier, I.: Three-dimensional object recognition is viewpoint dependent. Nat. Neurosci. 1(4), 275–277 (1998)

[231] Thomas, A., Ferrari, V., Leibe, B., Tuytelaars, T., Schiele, B., Van Gool, L.: Towards multi-view object class detection. In: Proceedings of the IEEE Conf. on Computer Vision and Pattern Recognition, New York, United States, vol. 2, p. 1589 (2006)

[232] Thrun, S., Fox, D., Burgardt, W.: Probabilistic Robotics (Intelligent Robotics and Autonomous Agents). The MIT Press (September 2005)

[233] Todorovic, S., Ahuja, N.: Unsupervised category modeling, recognition, and segmentation in images. IEEE Transactions on Pattern Analysis and Machine Intelligence 30(12), 2158–2174 (2008)

[234] Torralba, A., Murphy, K.P., Freeman, W.T.: Sharing visual features for multiclass and multiview object detection. IEEE Transactions on Pattern Analysis and Machine Intelligence 29(5), 854–869 (2007)

[235] Tuytelaars, T., Mikolajczyk, K.: Local Invariant Feature Detectors: A Survey. Now Publishers Inc., Hanover (2008)

[236] Veit, T., Tarel, J.-P., Nicolle, P., Charbonnier, P.: Evaluation of road marking feature extraction. In: Proceedings of 11th IEEE Conf. on Intelligent Transportation Systems, Beijing, China, pp. 174–181 (2008)

[237] Viola, P., Jones, M.: Robust real-time face detection. International Journal of Computer Vision 57, 137–154 (2004)

[238] Wachtert, S., Nageltt, H.-H.: Tracking of persons in monocular image sequences. In: IEEE Workshop on Motion of Non-Rigid and Articulated Objects, Washington, DC, USA, p. 2 (1997)

[239] Wahl, E., Strobel, T., Russ, A., Rossberg, D., Therburg, R.-D.: A motion-stereo based parking assistant. In: 16. Aachener Kolloquium Fahrzeug- und Motorentechnik (2007)

[240] Wahl, F.M.: Digitale Bildsignalverarbeitung: Grundlagen, Verfahren, Beispiele. Springer, Heidelberg (1989)

[241] Wahl, F.M., Biland, H.-P.: Decomposition of polyhedral scenes in hough space. In: Proceedings of the 8th International Conf. on Pattern Recognition, Paris, pp. 78–84 (1986)

[242] Wang, L., Tan, T., Member, S., Ning, H., Hu, W.: Silhouette analysis-based gait recognition for human identification. IEEE Transactions on Pattern Analysis and Machine Intelligence 25, 1505–1518 (2003)

[243] Wang, X., Ma, X., Grimson, W.: Unsupervised activity perception in crowded and complicated scenes using hierarchical bayesian models. IEEE Transactions on Pattern Analysis and Machine Intelligence 31(3), 539–555 (2009)

[244] Wang, Y., Tran, D., Liao, Z.: Learning hierarchical poselets for human parsing. In: Proceedings of the IEEE Conf. on Computer Vision and Pattern Recognition, pp. 1705–1712. IEEE (2011)

[245] Weber, M., Welling, M., Perona, P.: Towards automatic discovery of object categories. In: Proceedings of the IEEE Conf. on Computer Vision and Pattern Recognition, pp. 101–108 (2000)

[246] Weber, M., Welling, M., Perona, P.: Unsupervised learning of models for recognition. In: Vernon, D. (ed.) ECCV 2000. LNCS, vol. 1842, pp. 18–32. Springer, Heidelberg (2000)

[247] Wettler, M., Rapp, R.: Computation of word associations based on the co-occurrences of words in large corpora. In: Proceedings of the 1st Workshop on Very Large Corpora: Academic and Industrial Perspectives, pp. 84–93 (1993)

[248] Wiliem, A., Madasu, V., Boles, W., Yarlagadda, P.: A suspicious behaviour detection using a context space model for smart surveillance systems. Comput. Vis. Image Underst. 116(2), 194–209 (2012)

[249] Winn, J., Criminisi, A., Minka, T.: Object categorization by learned universal visual dictionary. In: Proceedings of the IEEE International Conf. on Computer Vision, pp. 1800–1807 (2005)

[250] Wiskott, L., Sejnowski, T.J.: Slow feature analysis: Unsupervised learning of invariances. Neural Computation 14, 715–770 (2002)

[251] Wojek, C., Roth, S., Schindler, K., Schiele, B.: Monocular 3D scene modeling and inference: Understanding multi-object traffic scenes. In: Daniilidis, K., Maragos, P., Paragios, N. (eds.) ECCV 2010, Part IV. LNCS, vol. 6314, pp. 467–481. Springer, Heidelberg (2010)

[252] Wojek, C., Schiele, B.: A dynamic conditional random field model for joint labeling of object and scene classes. In: Forsyth, D., Torr, P., Zisserman, A. (eds.) ECCV 2008, Part IV. LNCS, vol. 5305, pp. 733–747. Springer, Heidelberg (2008)

[253] Won, W.-J., Lee, M., Son, J.-W.: Implementation of road traffic signs detection based on saliency map model. In: Proceedings of IEEE Intelligent Vehicles Symposium (2008)

[254] Wu, J., Osuntogun, A., Choudhury, T., Philipose, M., Rehg, J.M.: A scalable approach to activity recognition based on object use. In: Proceedings of the IEEE International Conf. on Computer Vision (2007)

[255] Xiang, T., Gong, S.: Incremental and adaptive abnormal behaviour detection. Computer Vision and Image Understanding 111(1), 59–73 (2008)

[256] Yam, C., Nixon, M., Carter, J.: On the relationship of human walking and running: automatic person identification by gait. In: International Conf. on Pattern Recognition, Quebec, Canada (2002)

[257] Yamato, J., Ohya, J., Ishii, K.: Recognizing human action in time-sequential images using hidden markovmodel. In: Proceedings of the IEEE Conf. on Computer Vision and Pattern Recognition, pp. 379–385 (1992)

[258] Yao, A., Gall, J., Gool, L.V.: A hough transform-based voting framework for action recognition. In: Proceedings of the IEEE Conf. on Computer Vision and Pattern Recognition, pp. 1–8 (2010)

[259] Yedidia, J.S., Freeman, W.T., Weiss, Y.: Constructing free energy approximations and generalized belief propagationalgorithms. IEEE Transactions on Information Theory 51, 2282–2312 (2005)

[260] Yu, S.X., Zhang, H., Malik, J.: Inferring spatial layout from a single image via depth-ordered grouping. In: Proceedings of the IEEE Conf. on Computer Vision and Pattern Recognition Workshop (2008)

[261] Yuille, A.L.: Deformable templates for face recognition. J. Cognitive Neuroscience 3(1), 59–70 (1991)

[262] Zach, C., Pock, T., Bischof, H.: A duality based approach for realtime TV-L^1 optical flow. In: Hamprecht, F.A., Schnörr, C., Jähne, B. (eds.) DAGM 2007. LNCS, vol. 4713, pp. 214–223. Springer, Heidelberg (2007)

[263] Zhang, J., Liu, Y., Luo, J., Collins, R.: Body localization in still images using hierarchical models and hybrid search. In: Proceedings of the IEEE Conf. on Computer Vision and Pattern Recognition, Washington, DC, USA, vol. 2, pp. 1536–1543 (2006)

[264] Zhang, Z., Tao, D.: Slow feature analysis for human action recognition. IEEE Transactions on Pattern Analysis and Machine Intelligence 34(3), 436–450 (2012)

[265] Zhao, G.-W., Yuta, S.: Obstacle detection by vision system for an autonomous vehicle. In: Proceedings of IEEE Intelligent Vehicles Symposium, pp. 31–36 (1993)

[266] Zhu, L., Chen, Y., Torralba, A., Freeman, W.T., Yuille, A.L.: Part and appearance sharing: Recursive compositional models for multi-view. In: Proceedings of the IEEE Conf. on Computer Vision and Pattern Recognition, pp. 1919–1926 (2010)

[267] Zhu, L., Chen, Y., Yuille, A.: Unsupervised learning of a probabilistic grammar for object detection and parsing. In: Advances in Neural Information Processing Systems 19. MIT Press (2007)

[268] Zhu, L., Lin, C., Huang, H., Chen, Y., Yuille, A.L.: Unsupervised structure learning: Hierarchical recursive composition, suspicious coincidence and competitive exclusion. In: Forsyth, D., Torr, P., Zisserman, A. (eds.) ECCV 2008, Part II. LNCS, vol. 5303, pp. 759–773. Springer, Heidelberg (2008)

[269] Zhu, S.-C., Mumford, D.: A stochastic grammar of images. Found. Trends. Comput. Graph. Vis. 2(4), 259–362 (2006)

Printed by Printforce, the Netherlands